MW00716634

More Country Chuckles, Cracks & Knee-Slappers

Edited by Sandra Lessiter

Lessiter Publications, Brookfield, Wis.

Publisher's Cataloging-in-Publication
(Provided by Quality Books, Inc.)

More country chuckles, cracks & knee-slappers / edited by
Sandra Lessiter. -- 1st ed.
p. cm
Includes index.
ISBN: 0-944079-30-X

1. Wit and humor--United States. 2 Country life--
United States. I. Lessiter, Sandra Ann. II. Title:
More country chuckles, cracks & knee-slappers

PN6157.M67 1998 818'.02
QBI98-1274

International Standard Book Number: O-944079-30-X

Copyright© by Lessiter Publications

Cover design and interior art: Greg Kot

All rights reserved. No part of this book may be reproduced
in any form without the written consent of the publisher.

Published by Lessiter Publications,
P.O. Box 624, Brookfield, Wisconsin 53008-0624.
Telephone: (800) 645-8455 in the U.S. or (414) 782-4480. Fax: (414) 782-1252.
E-mail: info@lesspub.com
Manufactured in the United States of America

Dedicated to Janet Roberts...
a farm woman
who usually had a smile
on her face and always
enjoyed a good laugh.

Country Humor... Spreading It Around

YOU MAY HAVE once read the words of Mary Wilson Little, who said: "A good laugh is like manure to a farmer—it doesn't do any good until you spread it around."

Evident from the success of the first *Country Chuckles, Cracks & Knee-Slappers* book, we know that humor is a well-entrenched virtue of country life, and that farmers like to "spread" laughs whenever presented with the chance.

So when readers clamored for a "volume-two country joke book," my mother- and father-in-law, Pam and Frank Lessiter, approached me about compiling the second country joke book, just as my husband previously did with the first book.

I can tell this about how my "job" went...More than once our lazy hound dog awakened due to my laughter at the one-liners and funny situations painted in the following pages. I doubt that I ever spent as much time fighting back the giggles as I did in putting these pages together.

So what you have in front of you today is another brand new source of farm humor. This book contains 1,087 additional jokes that have appeared in the pages of *Farmer's Digest* over the last 60-plus years, as well as more recent submissions from farmers who scribbled a note to us saying, *"Here's another good one for your next joke book."*

A special thanks goes out to those loyal readers of *Farmer's Digest* who have shared a lifetime of their favorite jokes and stories with the farming world. Also, to the readers of the first *Country Chuckles, Cracks & Knee-Slappers,* who made this "offspring book" that you now hold in front of you possible.

The farm community is a unique one in that getting your co-worker, neighbor, supplier, spouse, son or daughter to let out a good hearty laugh is as much a part of the daily lifestyle as heading out to the barn for a full day of chores.

Country folks can laugh at themselves better than most and, in the process, make their own lives and that of those around them far more enjoyable. Whatever happens to the ever-evolving definition of "country life," those joke-trading skills and the ability to laugh at whatever predicament is thrown their way will never die.

I hope you enjoy this book as much as we did in putting it together for you. And do what Mary Wilson Little suggested—you know, remember to "spread" your humor around. Happy laughing!

—Sandy Lessiter

Three Favorites From America's Worst Farmer...

• A government farm expert dropped by Ory Klickenbacker's place one day to offer Ory some advice on his farm plan. He said, "Ory, you need to diversify. Get out of the corn, soybeans and wheat rut and plant some of the alternate, higher profit, less labor intensive crops."

Ory decided maybe the guy was right, so he tried one. But just before he made his first harvest...the sheriff found out about it and came out and cut it all down!

• Ory was on the back of the farm one day, just sitting there enjoying a nice autumn day, lost in his thoughts, watching the cattle graze.

He sat there watching the cattle for an hour or more when it finally dawned on him, "These are not my cattle!"

• It was mid-July and Ory still hadn't planted his corn. When George asked him about it, Ory drawled, "Well, when I was a young man my old Daddy always told me, 'Son, it's not time to plant corn until the hickory leaves get as big as a squirrel's ear.'

"By the time I figured out that the hickory tree I was watching was dead, it was too late to plant corn!"

You Deserve A Good Belly Laugh

IF THERE'S one thing farmers need more than rain and $5 per bushel corn, it's a good laugh. You're holding in your hand the best source of laughs for farmers I've come across in a long time. *More Country Chuckles, Cracks & Knee-Slappers* starts off with 1,087 exciting new jokes and stories to go along with the original *County Chuckles, Cracks & Knee-Slappers* book that featured 1,241 of the best country jokes you'll find anywhere.

As a speaker who has spoken at hundreds of farm meetings over the past several years, I can attest to the value of this book. In fact, I wouldn't have an act at all if it weren't for stealing jokes from the first *Chuckles* book have an act at all if it weren't for stealing jokes from the first *Chuckles* book and I can't wait to try out some of the hilarious material I've already read in this brand new volume of country fun.

But you don't have to be a farmer to get a kick out of the funnies contained in this book. Anyone who enjoys a good laugh will not be able to put this book down.

Here's a suggestion for using this book. Pick out a couple of your favorite stories or jokes, memorize them and practice telling them to the spouse, the kids, the livestock, whoever. Try them out on a couple of friends. Really learn how to tell them.

Then when you get in a crowd, just nonchalantly lay one of your favorite funnies from this book on them. After a while, spring another one on them. They'll walk away muttering to one another, "Where does that guy come up with all that stuff?"

The more of these anecdotes you know, the bigger hit you'll be. But let me caution you. Don't try to tell so many that you can't do a good job on them. A poorly told joke is worse than no joke at all. Just take it slow and easy. Learn a couple jokes, then go back to the book and pick out a couple more. Before long, you'll have a fair-sized repertoire *(now there's a real fancy word you don't hear much around the livestock sale barn)* of funnies.

Practice until you've got them down pat and you're confident in your ability to tell them. You'll love the results! There's no doubt that you could put together an excellent monologue right out of this book. But don't get too good at it because I'm out there in rural American trying to make a living telling stories like these to fellow farmers and ag suppliers at meetings and conferences.

Here's another idea that I use. Keep a pencil or colored highlighter handy when you're reading this book for the first time. When you come upon a joke or story that you especially like, mark it. Then someday when you need a good laugh in a hurry, all you have to do is pick up this book and scan through its pages to quickly find one of your favorites.

So take this book and get in a room by yourself. (but stay close enough that you can holler, *"Honey listen to this..."*) turn off the television, get a cold soft drink and you'll have yourself a big time! Just you and this brand new *More Country Chuckles, Cracks & Knee-Slappers!*

—Lewis Baumgartner
World's Worst Farmer
Fulton, Missouri

Okay, Let's Get Started!

While in the big city on banking business, a farmer walked into a restaurant and asked for a table. He was stopped by the maitre d' for not wearing a tie.

Frustrated, tired and hungry after driving 125 miles into the city, the farmer returned to his pickup truck and dug around behind the seat. Yet he couldn't find anything suitable to wear as a necktie. Then he spotted his jumper cables, draped them around his neck, fashioned them into a crude bow tie and returned to the restaurant eager to sit down to a big meal.

The maitre d' looked him over and smirked, "Well, you're resourceful if nothing else," he said after a moment. "Come on in—but don't start anything."

Where You'll Find These Country Funnies

Chapter 1

Working Hard... Or Hardly Working?

"Hard work never killed anybody, but it has been known to scare some people half to death."

● The farm worker approached his boss, "I'd like to have next Wednesday off, sir."

The boss responded, "Why?"

"It's our silver anniversary," replied the worker, "and my wife and I want to go out and celebrate."

The boss asked, "Are we going to have to put up with this every 25 years?"

● The new farm hand limped up to the boss at the end of a long day of backbreaking work.

"Boss, are you sure you got my name right?" he asked.

"It's right here—you're Joe Simpson, aren't you?" the boss replied.

"Yeah, that's it," moaned the fellow. "I was just checking. I thought maybe you had me down as Samson."

● After the hired hand made a foolish and costly error on the farm, he said to his boss, "I suppose you think I'm the perfect idiot."

The farmer answered, "Not at all—nobody's perfect."

● "Working here builds character," the farm boss told his newest hired hand. "If you don't believe me," he continued, "just look at the characters who work here."

● Farmer Dan was cussing out his new farm hand. "You're the slowest man I've ever seen! Don't you do anything quickly?" he yelled.

"Well," said the hired hand in between yawns, "I do get tired fast."

● A farm hand was handed a pay envelope which, by error, contained a blank check.

He looked at it and moaned, "Just what I thought would happen—my deductions finally caught up with my salary."

● The job applicant was asking the farmer some questions about the position. "Why did your ad say you wanted to hire a married man?" he inquired.

"They don't get so upset when I yell at 'em," the farmer replied.

● "Before I take this job for the summer," the teenager said, "tell me, are the hours long?"

"No," said the farmer, "only the usual 60 minutes each."

● A farmer was asked by a hired hand he had fired for a recommendation letter. He thought it over and then wrote: "The bearer of this letter is leaving me after one month's work. I am perfectly satisfied."

● A farmer and his farm hand were having a rare conversation about business. "Boss, there have been so many people replaced by machines, I'm afraid I may be next," lamented the farm hand.

"Don't worry," the farmer said. "They haven't invented a machine yet that does nothing."

● Two teenagers were talking as they walked down the street. "So, your new job at Smith's farm makes you pretty independent, huh?" Bill asked Pat.

"Absolutely," declared Pat. "I get there any time I want before 7 a.m. and leave just whenever I please after seven at night."

● "I'm sorry we can't hire you," the farmer told the young man, "but there's just not enough to keep you busy."

"Sir," the man persisted, "you'd be surprised at how little it takes."

● "Your references are good; I'll try you," said the farmer to the young man, applying for the job on the dairy farm.

"Wonderful," replied the young man. "Is there any chance to rise, sir?"

"I'll say there is," said the farmer. "You'll rise at a quarter to five every morning."

● Ad in a farm paper: "Wanted—dairy farm employee. Must not have any bad habits such as drinking, cussing or eating margarine."

● A farmer happened across some hunters, walking closely to his land.

"Say!" the farmer called, "don't shoot anything that isn't moving. It may be my hired man."

● Lousy Lenny came in to work late again one morning. "Sorry," he apologized. "I had car trouble."

Lenny's employer looked at him suspiciously. "What happened to your car?" the farmer asked.

"Uh," stumbled the farm hand, "I was late getting into it."

● A cameraman, working for the educational department of a film company, met an old farmer in town. Putting it into terms the old man could understand, he explained to the farmer, "I've been taking some moving pictures of life out on your farm."

"Did you catch any of my men in motion?" asked the old farmer curiously.

"Sure I did."

The farmer shook his head reflectively, then commented, "Modern technology is a wonderful thing."

● Tardy Tim was late again.

"You should have been here at 7:30!" the farmer scolded.

"Why?" Tim said sleepily, "What happened?"

● A hired hand received his paycheck and upon inspecting it, found it to be $10 too much. He decided not to say anything to the boss.

When he got his next check, it was $2 too small. He immediately went to talk to the boss.

The boss argued, "We have been checking our records and found out we paid you $10 too much last month. How come you didn't say anything then?"

To which the hired man replied, "Well, anybody can make one mistake, but when he makes the second, that is just too much."

● After the farm hand was fired and told to move his family out of the hired hand's house, the wife walked into a pet store and asked the owner for 200 cockroaches. The clerk considered this to be an odd request and asked the customer why she needed so many.

"Well, we've been evicted from the Perkins' farm," she said, "and they told us to leave the house in the same condition as when we moved in."

● The county census-taker knocked on the farm house door and Farmer Ned answered. "How many people work here on your farm, sir?" the census-taker asked.

Ned replied, "Oh, I'd say about half."

● *Farm Employee's 10 Commandments...*

1 If at first you don't succeed, destroy all evidence that you've tried.
2 A conclusion is the place where you got tired of thinking.
3 Experience is something you don't get until just after you need it.
4 For every action, there is an equal and opposite criticism.
5 He who hesitates is probably right.
6 No one is listening until you make a mistake.
7 Success always occurs in private, and failure in full view.
8 To steal ideas from one person is plagiarism; to steal from many is research.
9 The sooner you fall behind, the more time you'll have to catch up.
10 Work is accomplished by those employees who are still striving to reach their level of incompetence.

● Farmer Vern was interviewing a hired hand. "You say you were at your last place for 23 years? And you left the place after putting in that much time?"

The prospective employee answered, "Well, sir, that's what happens when you're paroled."

● A young man had just been fired as one of Farmer Bert's farm hands. As he was picking up his final check, he ran into his boss.

"In a way, I'm sorry to lose you. You've been just like a son to me," Bert told the departing employee.

"Really, sir?" the ex-farm hand replied.

"Yup, just like him: insolent, surly and unappreciative."

● Farmer Don approached his hired hand, who was resting on a tree stump. "Smith," he said, "we're giving you a raise."

"Why, thanks, sir," the employee said.

"Yup, we want your last week here to be a happy one."

● A somewhat slow farm hand was helping his boss trim trees around the farm. He was concentrating on the task at hand, when suddenly the saw slipped and cut off his ear!

He immediately climbed down from the tree and was searching for it on the ground when the farmer came by and asked him what he was looking for.

"I'm looking for my ear," the farm hand answered.

"Well, there's an ear over here," his boss said, pointing to a flesh-colored object in the soil. "That must be yours."

"No," came the reply. "Mine had a pencil behind it."

Chapter 2

It's A Living

"Maybe they call it take-home pay because there is no other place you can afford to go with it."

—Franklin P. Jones

● When the plumber arrived late, he inquired, "How have you managed?"

The wife replied, "Not too badly—while we were waiting for you to get here, I taught the children how to swim."

● A farm woman, who'd just finished a first aid course, spied a man lying face down in a flooded gutter. She rushed over to him, turned him on his back and began to apply mouth-to-mouth resuscitation.

Suddenly, the man sat up and pushed the woman away. "I don't know what you got on your mind, lady," he said, "but I need to get back to clearing out this drain."

● The pope has the best job in the world: he has only one boss, and he only meets him after he dies.

● A professional carpet-layer stepped back to survey a newly installed carpet. Reaching into his shirt pocket for a cigarette, he realized the pack was missing. At the same time, he noticed a lump under the carpet in the middle of the room, about the size of the missing cigarette pack.

There was no way to retrieve his pack from under the attached carpet without ripping everything up and starting over. Finally, he decided to beat the object flat, thereby destroying any evidence of his mistake.

Satisfied with the job, he gathered his tools and walked out to his truck. There on the seat of the truck was the mislaid pack of cigarettes. As he lit one up, the homeowner hurried out of the house and asked, "Hey, have you seen my son's gerbil?"

● Lenny, the owner of a hair salon in a small town, enjoyed the peace of mind that his was the only salon in town. He was responsible for cutting and styling the hair of just about everyone who lived there. Lenny's income allowed him to live comfortably and even send all six of his children to college.

Unfortunately for Lenny, big business came to his town. Right across the street from his busy little hair salon sprang up one of those new full-service salon franchises.

Immediately, the media campaign began: ads in newspapers, magazines and billboards announced, "Everything for $6! $6 haircuts, $6 perms, $6 manicures, everything for $6!"

Soon all of Lenny's customers started dropping off, and his business sat empty. Lenny, however, was not a quitter. He called upon his best friend who had experience in marketing and together they came up with an idea that would ultimately save Lenny's business. They painted a sign that said, "We Fix $6 Haircuts!"

● "The sausages you sold me have meat at one end and bread crumbs at the other, " stormed the lady customer.

"Yes, madam," replied the butcher, "in these tough times, it's pretty hard to make both ends meat."

● The local garage in a small Kansas country town caught fire and was totally enveloped in flames. The local fire engine with six volunteers clinging to its sides came charging up to the fire at high speed. The fire wagon drove right into the garage and the middle of the flames. The six fire fighters jumped off and bravely put out the fire in no time flat.

A tourist from the city, who happened to be wealthy, was so impressed with their brave and efficient work that he donated $500 to the fire department to use for equipment. He asked them, "What will you use the money for?"

"First," drawled the fire chief, "we're going to fix the brakes on the fire truck."

● A new barber nicked a customer badly while giving him a shave. Hoping to restore the man's feeling of well-being, he asked solicitously, "Do you want your head wrapped in a hot towel?"

"No, thanks," said the customer, "I'll carry it home under my arm."

● Sam the barber seemed jumpy and it made his customer nervous. "Sam," he said, "what happens if you cut a customer? Does the boss get angry?"

"Sure, he does," replied Sam. "He makes me pay a dollar for every cut we give a customer... but I don't care. I had a good day at the races yesterday."

● Mrs. Smith was angry with her grocer. "I sent my son for two pounds of peanuts, and you sent me only one and a half pounds."

"Ma'am, my scales are correct," the grocer said defensively. "Have you weighed your son?"

● Two mechanics were talking shop.
"Which do you prefer for upholstery in cars?" asked the first. "Leather or fabric?"
"I like fabrics," replied the second. "Leather is too hard to wipe your hands on."

● The farmer was closing out his bill at the local restaurant.
"Waiter," he said, "I find that I have just enough money to pay for the dinner, but I have nothing in the way of a tip for you."
The waiter thought for a moment and said, "Hmm, let me add up that bill again, sir."

● A nervous little man walked into a grocery store in a small Kansas town.
"I want to buy all your over-ripe vegetables and stale eggs," he said.
"Well," said the shopkeeper with a twinkle in his eye, "you must be going to see the new comedian at the theater tonight."
"Not so loud," said the man, looking around apprehensively. "I am the new comedian!"

● Two farmers sat over their coffee for two hours, taking up space in a busy eatery. Finally, the waitress handed them a check for $10.
"Wow! Is that for coffee?" one of them asked.
"No," replied the waitress. "It's a parking ticket."

● The new secretary ignored the telephone when it rang. The irate boss finally told her to answer it every time it rang.
She retorted, "Now, that is ridiculous. We both know it's for you; none of my friends know I'm here yet."

● Two seed dealers were talking to each other about the ideal job. "Isn't there a job out there where you get a ton of money, but someone else does the work and takes the blame when things go wrong?"
A receptionist, working in the same office, overheard the two dealers, and piped up, "Oh yeah, there are thousands of those jobs. Ask any secretary about her bosses."

● The veterinarian stopped at the local diner for a quick break between making farm calls. But when his order came up, he looked disturbed. "Why is my doughnut all squished up?" he asked the waiter.
The waiter looked at him strangely and said, "Well, you said, 'I want a

cup of coffee and a doughnut and step on it!'"

● A woman rushed into a country department store to buy a package of invisible hairpins.

"Are you absolutely sure these are invisible?" she asked the clerk.

After he explained that they indeed were, she still didn't believe him.

"Listen, lady," said the country boy, growing impatient, "I've sold $20 worth this morning, and we've been out of them for three weeks."

● The other morning a postman limped into the post office. His pant leg was torn and he looked mad. The postmaster asked him what happened. "I was attacked by a large, brown dog that bit me in the calf," the postman explained angrily.

Walking over and trying to help, the postmaster asked, "Did you put something on it?"

"No," grumbled the postman, "the dog seemed to like it plain."

● The wife of a farmer frequently sold her surplus butter to a grocer. On one occasion, the grocer said to her, "Excuse me, ma'am, but your pound of butter was underweight last week."

"Now fancy that!" said Mrs. Farmer. "My baby was playing with the weight that day. When I couldn't find it, I used the pound of sugar you sold me!"

● "It's tough," said the woman to the butcher, "to pay $2.50 for a pound of hamburger."

"Well," said the butcher, "we have tougher for $2 a pound."

● A country store meat counter clerk, who had a particularly good day, proudly flipped his last chicken onto the scale and weighed it. "This will be $7.70," he told the customer.

"That's really a little too small," said the woman. "Don't you have anything larger?"

Hesitating, but thinking fast, the clerk returned the chicken to the refrigerator, paused a moment, then took it out again.

"This one," he said, "will be $8.50."

The woman paused a moment, then made her decision. "I know how to solve this," she said. "I'll take both."

● A farm boy sacking groceries at the town grocery store watched with fascination as the store installed a huge new machine for squeezing fresh orange juice. The country boy was captivated by the contraption and begged to be trained to use it.

"Sorry, kid," the store manager said. "Baggers can't be juicers."

● A farmer asked one applicant whether he had any experience as a lumberjack. "You kidding?" he answered. "I used to be foreman at the Sahara Forest."

"You must mean the Sahara Desert," replied the interviewer.

"Well, yeah," the lumberjack shrugged, "now."

● The used car salesman, having made a quick sale, was alarmed to see the purchaser driving back into the lot.

"Nothing wrong, is there?" the salesman said with a grin.

"No, not a thing," the customer replied. "But the little old lady you said owned the car before left some things in it. Here's her pipe and tobacco pouch that were in the glove compartment and the bottle of scotch she forgot to take out from under the seat."

● After years of being blasted into a net, the human cannonball went to the circus owner and told him he was going to retire to his farm in the country.

"But you can't retire!" shouted the cigar-chomping boss. "Where am I going to find a man of your caliber?"

● A down-on-his-luck painter got the contract to paint the rural church. In order to make more money, he thinned the paint, thinking no one would ever be the wiser. But just as he finished the job, there was a heavy rainstorm, which washed off all of the paint he had just applied.

Before he could do anything, out of the clouds came a loud voice: "Repaint! Repaint, my son, and thin no more!"

● "Don't feel any pressure to make a decision today," said the slippery life insurance salesman to the farmer. "Sleep on it, and if you wake up in the morning, you can give me a call."

● A young businessman was leaving his office at the end of the day when he found his boss standing in front of a shredder with a piece of paper in his hand. "Listen," said the boss, "this is a very sensitive and important document, and my secretary has left already. Can you make this thing work?"

"Of course," said the young man, turning the machine on, inserting the paper and pressing the start button.

"Excellent, excellent!" said the boss as his paper disappeared inside the machine. "I just need one copy."

What Your Boss Really Means On Your Farm Job Performance Evaluation...

Accepts new work assignments enthusiastically: Never completes a task.
Active socially: Drinks a lot.
Alert to company developments: Gossips.
Approaches difficult problems logically: Finds someone else to do the job.
Average: Not too bright.
Character above reproach: Still one step ahead of the law.
Charismatic: Not open to the opinions of others.
Competent: Able to complete work if supervisor helps.
Consults with co-workers often: Indecisive, confused and clueless.
Consults with supervisor often: Pain in the butt.
Delegates responsibility well: Knows how to pass the buck.
Demonstrates leadership qualities: Has loud voice.
Deserves promotion: Make up new title to make him feel appreciated.
Displays excellent intuitive judgment: Knows when to disappear.
Displays great dexterity and agility: Dodges and evades superiors well.
Enjoys job: Needs more to do.
Excels in sustaining concentration but avoids confrontations: Ignores everyone.
Applies skills well: Makes good cup of coffee.
Expresses self well: Can string two sentences together.
Happy: Paid too much.
Hard worker: Usually does it the hard way.
Identifies major management problems: Complains a lot.
Is well informed: Knows all office gossip and where all the skeletons are kept.
Is unusually loyal: Wanted by no one else.

Chapter 3

Enjoying The Off Hours

"A bad day of fishing is still better than a good day of farming."

● Two farmers had been out in the woods hunting for several hours. One of them had been growing uneasy. Finally, panic overcame him. "We're lost!" he cried to his companion. "What on earth are we gonna do?"

"Take it easy," said his composed friend. "Shoot an extra deer and the game warden will be here before you know it."

● Two drunks wandered into a city zoo and as they staggered past a lion's cage, the king of beasts let out a terrific roar.

"C'mon, let's get out of here," said the first drunk.

"You go ahead if you want to," replied his more inebriated cohort. "I'm gonna stay here for the movie!"

● The stingy farm loan officer left a plush night club one night, walking past the doorman without tipping him. Nevertheless, the doorman helped the man into the car and said pleasantly, "In case you happen to lose your wallet on the way home, sir, just remember that you didn't pull it out here."

● "Why don't you play golf with Harry anymore?" the farm wife asked her husband.

"Would you play with someone who cheated all the time?" her husband asked.

"I guess not," she answered.

He added, "Well, neither will Harry."

● A farmer who had to do a lot of traveling decided to buy his own plane and take flying lessons. He became a good pilot and a couple of years later, traded in the plane for one with pontoons, since he had a vacation home on a beautiful lake tucked away in the wilderness.

On his first trip in the new plane, however, he headed for a landing at the airport, forgetting that his new plane landed only on water. Fortunately his wife was with him, and when she realized what he was doing, she shouted, "Pull up! Pull up! You can't land this thing on a runway! You don't have any wheels; you've got pontoons!"

Looking sheepish and embarrassed, the farmer hit the throttle and got out of the landing pattern. He landed safely on the lake, and with a sigh of relief, he shook his head and said to his wife, "I don't know what I was thinking. That was the dumbest thing I've ever done."

Then he opened the plane door and stepped out into the lake.

● Roy and Ted, two simple-minded farm hands, decided to go camping. After they attached the trailer to the car, Roy wanted to make sure that the car was in good condition before they took off. So, he asked Ted to go in front of the car to check the headlights.

Roy switched the headlights on. "Yeah, they're workin'," reported Ted.

Then Roy turned on the high beams. "Check!" Ted called out.

Roy then asked Ted to go to the rear of the car to check the brake lights. Roy slammed on the brake, and Ted yelled, "Yeah, they're fine!"

Last thing to check were the turn signals. As Ted stood by the left indicator, Roy switched on the turn signal. Ted was ready with his report, "It's working! Oops! It's not working! ...Wait! It's working! Oops! It's not working!..."

● Farmer Ray spent a rare day off down at the lake fishing. Disgruntled after a long day with no luck, he stopped in town at a fish market before returning home.

Shaking his head sadly, he instructed the clerk, "I need three of those big fish. And could you do me a favor? Before you wrap 'em up, toss 'em to me. That way, I won't be lying when I tell my wife I caught 'em."

● "Did you fish with flies?" a friend asked the farm supply clerk, who was just returning from vacation.

"Fish with' em?" the clerk said. "We fished with 'em, we ate with 'em, we slept with 'em..."

● "I've noticed," said the tourist to the fisherman, "that the size of your fish changes every time you tell that story to someone new."

"Of course," said the fisherman. "I never tell people more than they'll believe."

● "This food is terrible," Farmer Sam complained in the dining room of the Luxury Vacation Resort. "I want to see the owner of this restaurant!"

"Sorry, sir," the waiter replied. "He's out to lunch."

• In a saloon in the wild west, guns were blazing and men were running in all directions for cover, when a mild-looking man strolled in and leaned against the bar. In a little while, all was quiet.

A stranger came up and congratulated the man on his perfect nerve control. "Oh, that's nothin'," said the man with a smile. "I'm quite safe. You see, I owe money to everybody in this place."

• A pirate walked into a tavern after many weeks at sea. He had a wooden leg, a hook where his hand used to be and a patch over his left eye. The curious young man sitting next to him asked the pirate how he came to have a wooden leg.

The pirate replied, "Well, I was standin' on the deck of me ship one day when a wave washed me overboard. Then a shark came and bit me leg off."

The young man presses on, "How did you lose your hand?"

"Many years ago, I was fightin' the British, and one of those dastards cut me hand off," answered the pirate. "They couldn't get me another hand, so I got stuck with this here hook."

Next, the young man asked, "How did you lose your eye?"

"I was standin' on the deck of me boat," said the pirate, "and a bird flew over and *&%$@ in me eye."

The young man became very confused. "That's it? A bird *&%$@ in your eye and you lost it?"

The pirate replied, "Well, it was me first day with the hook."

• The farmer, out for a day of golf, was thoughtful as he approached the tee on a hole with a number of traps and a pond. He looked up at the sky and said, "Hmm, should I use a new ball or an old one?"

Much to the farmer's surprise, God answered. "Use a new ball," He advised.

The farmer teed up and heard God say, "Take a practice swing."

So, the farmer took a practice swing, and God said, "Take another practice swing."

The farmer took another practice swing.

After a long pause, God replied, "Use an old ball."

• "Don't you see that sign?" asked the farmer, pointing to a "No Fishing Here" notice tacked on a tree.

"Yes, I do," replied the disgusted angler, "and let me tell you, the fellow who put that sign up knew what he was talking about."

• Every baseball team could use a man who plays every position superbly, never strikes out and never makes an error.

The trouble is, there's no way to make him put down his beer and come out of the grandstand.

● "I figure," Farmer Derek said, "since three-fourths of the Earth's surface is water, and one-fourth is land, it's obvious that the Good Lord intended that man spend three times as much time fishin' as plowin'."

● Farmer George went out with the boys one evening, and before he realized it, the morning of the next day had dawned. He hesitated to call home and talk to his wife.

Finally, he hit upon an idea. He telephoned and when his wife answered he shouted, "Don't pay the ransom, honey! I escaped."

● A man was observing a farmer fishing for nearly two hours. The farmer, feeling a little uneasy by this, said, "Say, you've been standin' there watching me for two hours now. Why don't you try fishing yourself?"

"Me?" asked the spectator. "I don't think so. I ain't got the patience."

● It was a Sunday afternoon and a farmer had been watching football on TV, one game after another. Finally, he fell asleep in the chair and slept there all night.

When his wife arose in the morning, she was afraid he'd be late for chores. "Get up, dear," she said, "it's ten to seven."

In an instant, he was awake. "In whose favor?" he shouted.

● A fisherman who had been caught by the game warden for exceeding his limit on black bass was taken to the local courthouse, where he readily admitted his guilt.

"That'll be $25 per fish and court costs," pronounced the judge.

After paying the fine, the angler approached the judge and cheerfully asked, "Your Honor, if I may, I'd like several copies of the court record to show my friends."

● A distinguished-looking fellow was entranced with a model train that whistled, belched smoke, deposited milk cans, delivered mail and did practically everything that a real freight train does. He finally said, "I'll take it!"

"A wonderful choice!" exclaimed the approving clerk. "I'm sure your grandson will love it."

"Why, you're absolutely right," said the old gentleman. "I'd better take two."

● "I'd move heaven and earth to break 100," announced the disgruntled farmer as he banged away in a sand trap.

"Try heaven," advised his partner. "I think you've already moved enough earth."

● The hard-working veterinarian was spending an evening at home when a colleague called and invited him to a poker game at still another vet's home.

As he put on his coat, his wife sighed, "Do you have to go out tonight, dear?"

"Yes," he replied, "and it sounds serious...there are two other veterinarians there already."

● A man sat hunched over the bar with several empty glasses in front of him, patiently trying to spear an olive in his martini glass with a toothpick.

After he'd been at it a few minutes with no success, the farmer sitting next to him grabbed a toothpick and speared the olive. He held it out to the poor man and said, "See how easy that was?"

"Sure," came the reply, "after I tired him out for you!"

● Two old farmers were playing cards at the barber shop.

"Hey, Ralph, do you know how to get three old ladies to say &%$#*!?" one asked.

Ralph shook his head and asked, "How?"

The first burst out, "You get a fourth lady to yell, 'Bingo!'"

● Two farm supply store workers decided to have a few drinks after work. After one too many, they left the bar and wandered out onto a railroad track.

"Man," the first one complained, "this a long flight of steps!"

"I don't mind the steps so much," replied his drunk co-worker. "It's the low railing that bothers me."

● A farm hand and his wife were about to go on a trip to the woods to camp for a week.

"Well, we're repairing to go to the woods," he told his friends at the local feed mill the day before his vacation.

"You mean preparing. Repairing means to fix something," the feed manager replied as he rolled his eyes.

"That's right," said the farm hand. "We're fixin' to go on the trip!"

● What's the difference between a bad golfer and a bad skydiver?

A bad golfer goes, "Whack!—#%&*!"
A bad skydiver goes, "#%&*—Whack!"

● "You should advertise!" the canvasser told the proprietor of a small country store.

"No, sir! I'm against it," came the emphatic answer.

"But why?"

"Advertising don't leave a man no time," the man replied. "I tried it once last year, and I was so doggone busy I couldn't get in no fishin' all summer!"

● Two Illinois hunters drove up to a farm in Northern Wisconsin. One stayed in the car while the other walked up to inquire at the farmhouse.

"Do you mind if we hunt on your land?" he asked the farmer.

The farmer responded, "You can hunt here, but do me a favor. My horse is very sick and must be put down. I just can't get myself to do it. Would you shoot it for me?"

The hunter sympathized, "I know how you feel. Sure, I'll shoot the horse for you." With that, the hunter returned to the car and decided to pull a prank on his pal. He said, "That farmer is a real jerk. He won't let us hunt here. I'm so mad, I'm going to shoot his horse." And he shot it.

Then he heard another shot from behind him. He turned and saw his friend holding a smoking rifle, "I got his bull, too," his friend said proudly.

COW HIDE

● A man was sprawled out across three seats in the fifth row of a movie theater.

"How rude of you, taking up three seats like that!" huffed a passing woman. "Didn't you learn any manners? Just where did you come from?"

The man pointed feebly up and gasped, "The balcony!"

● "Well, young man, how do you like my game?" the over-confident golfer asked.

"It's all right, sir, I suppose," the caddy replied, "but I still prefer golf."

● The golfer stormed off the 18th green, livid with rage. "That's it!" he yelled. "I'm through with this game forever," as he threw his bag of expensive clubs on the ground.

"If you want these clubs," he told the caddie, "they're all yours." The caddie accepted the clubs without a word.

The golfer then told the caddie he could have his bag, the cart, the umbrella and all the $5 balls.

"Gee, thanks a lot, mister," the caddie said.

As they started for the clubhouse, the golfer ripped off his shirt with the little alligator on it and threw it on the ground. He followed this by peeling off his slacks. "You can have those, too, if you want them," he said.

The caddie looked at the man standing in front of him in his shorts. "Sir, you can't walk into the clubhouse looking like that," he said. "They'll take away your membership."

"Who said anything about walking into the clubhouse?" the man answered. "I'm going to throw myself into this pond and drown!"

The caddie shook his head. "That's impossible, sir," he said.

"Why?" asked the irate golfer.

"Because," responded the caddie, "you just can't keep your head down."

● Two bar patrons were shooting the breeze over a couple of beers.

"Glasses do the strangest things to vision," remarked one as he shifted his specs on his nose.

"You're tellin' me," the other slurred, "especially after they've been filled a few times!"

● Golf is the game that turned the cows out of the pasture and let the bull in.

● A drunk got into a cab and told the cabbie, "Take me to Nolan's Bar and Grill."

"You're in front of Nolan's right now," said the cabdriver.

"OK," stammered the drunk, "but next time don't drive so fast."

● The tight-pursed Healys took their infant son to the movies. The usher warned them that if the child couldn't keep quiet, he would have to give their money back and ask them to leave the theater.

About half-way through the second feature, Healy nudged his wife and whispered, "What do you think?"

"This movie stinks," his wife replied.

"I agree," answered Healy. "Pinch the baby a few times."

● Under the influence of alcohol, a Frenchman wants to dance, a German to sing, a Spaniard to gamble, an Englishman to eat, an Italian to boast, a Russian to be affectionate, an Irishman to fight. And an American to make a speech.

● Two farmers were drinking in a bar, when suddenly, one of them fell flat on his face.

His buddy looked down and said, "That's what I like about you, Harvey, you know when to quit."

● Farmer Herb told his drinking buddy, who was sporting a black eye, "I think it was a pretty stupid move to call that guy's girlfriend ugly, Ralph."

"I didn't call her ugly," Ralph protested, getting back on his barstool, "I just asked if she was allowed on the furniture."

● Into a saloon came a fellow leading a horse. "Gimme a swig of whiskey for my friend here," he said to the bartender.

The barkeep obliged, and the fellow opened the horse's mouth, threw down the liquor and ordered another which went the same.

"What about yourself?" asked the bartender. "Ain'tcha gonna have anything?"

"Not me," said the horse's owner, mounting the animal and heading out of the saloon. "I'm driving."

Chapter 4

Home Sweet Country

"You can take the boy out of the country, but you can't take the country out of the boy."

—Arthur Baer

● Two small town McDonald's employees were commenting on the amount of vehicles in the parking lot. "Man, look at all those John Deeres here tonight," said one.

"Yeah," replied the other nostalgically. "Reminds me of my own prom night."

● A farmer walked up to the general delivery window at the post office where a new clerk was busy sorting letters.

"Any mail for Mike Howe?" the farmer requested loudly.

The clerk ignored him and the man repeated his question in an even louder voice. Without looking up, the clerk replied, "No! Not for your cow or your horse neither!"

● The hillbilly hadn't taken a bath for a long, long time. The situation grew so bad that his family finally put together a committee to force him to bathe.

The mountaineer objected fiercely. He kicked and he squawked, but finally they succeeded in undressing him.

After peeling off several garments, the man was down to his long flannel underwear. Under this, much to his surprise, was a sweater.

"Can yer imagine that?" he drawled. "And here I bin searchin' high and low fer that thar sweater fer over two years!"

● Making his first call after installing electricity in a backwoods cabin, the meter inspector found that very little current had been used. "Don't you ever use the light?"

"Sure," drawled the old fellow.

"How long at a time?" the inspector asked.

"Oh, just long enough to see to light the oil lamp."

LAMB CHOP

● A young hillbilly woman, away at the university, received the following letter from her ma...

"Dear Louanne Ellie Mae,

I'm writing this letter slow because I know you can't read fast. We don't live where we did when you left home. Your dad read in the paper that most accidents happen within 20 miles from home, so we moved.

I won't be able to send you the address because the last Arkansas family that lived here took the house numbers so that they wouldn't have to change their address.

This place is nice. It even has a washing machine. I'm not sure it works so good though. Last week I put a load in and pulled the chain and haven't seen it since.

The weather here isn't bad. It only rained twice last week; the first time for three days and the second time for four days.

About that coat you wanted me to send you, your Uncle Stanley said it would be too heavy to send in the mail with the buttons on, so we cut them off and put them in the pockets.

John locked his keys in the car yesterday. We were really worried because it took him two hours to get me and your father out.

Your sister had a baby this morning, but I haven't found out if it's a boy or a girl, so I don't know if you're an aunt or uncle."

—*Your mama*

● During a violent snowstorm, one Red Cross rescue team was finally transported by helicopter to within a mile of an endangered mountain cabin all but covered by deep snowdrifts.

The rescuers struggled on foot through the deep drifts and finally arrived at the cabin, where it took them three grueling hours to shovel away enough snow to clear the door. They knocked and a mountaineer answered. One rescuer stepped up and said,"We're here from the Red Cross."

"Well," said the mountaineer, closing the door slowly, "it's been a right tough winter. I just don't know how we can give this year."

● Harry Hillbilly's idea of saving money: Not working hard enough to need deodorant.

● In a corner of his log cabin, the mountaineer struggled with a pencil and a piece of paper. Suddenly, he looked up and exclaimed, "Doggone if I ain't learned to write!"

His wife walked over and peered at the lines scribbled on the paper. She asked, "What do it say, Ezra?"

"I don't know," he answered, "I ain't learned to read yet."

● General Custer rounded up his men for an impromptu meeting.

"Men, I've got some good news and some bad news," he told his troops. "The bad news is that the Indians have outnumbered us 30 to one. Looks like we're gonna get slaughtered."

"What's the good news, sir?" a voice from the back said.

"Well, at least we won't have to ride our horses back across Nebraska."

● A woman from Utah placed a call to her aunt, who lived in a small town in Iowa. After she gave her aunt's number to the Utah operator, the local Iowa operator came on the line.

"I'll ring her now," the operator informed the woman, "but I don't think they're home—their car is gone."

● The small Ohio town my uncle is from just named him fire chief. They had to; they're using his hose.

Three Strikes And You're Out

In the car on the way to the women's golf tournament, old Farmer Jethro was yelling at his wife, Gertrude, for everything under the sun—things like shopping too much and burning the toast at breakfast.

After driving off the first tee, Gertrude's ball sliced into the bushes. She rubbed the ball off and a genie appeared. "I am your servant, and I will grant you three wishes," the genie said. "However, whatever wish I grant you, your husband receives double."

For her first wish, Gertrude said she wanted to be a scratch golfer. "OK," said the genie, "but your husband is now the best golfer in the world."

For her second wish, Gertrude asked for $50 million. "OK," said the genie, "your wish is granted. But your husband now has $100 million."

Carefully contemplating her third wish, Gertrude suddenly smiled and said, "For my third wish, I'd like you to give me a mild heart attack!"

Chapter 5

Say Goodbye To The City

"There isn't much to be seen in a little town, but what you hear makes up for it."

—Frank Hubbard

● A little city boy spent his first night at the farm. Much earlier than usual, he was awakened by the activity around him.

Trudging down to the kitchen, where his grandmother was already busy cooking breakfast, the boy remarked sleepily, "I'll tell you. It doesn't take long to stay here all night, does it?"

● A farmer was giving his cousin Sally, visiting from the city, last-minute instructions before heading into town for the day.

"That guy from Sematol will be along today to inseminate one of the cows. I've hung a nail by the right stall so's you know which cow is to be impregnated."

Satisfied that he gave clear and ample instructions for his cousin, the farmer left to run his errands.

That afternoon, the inseminator arrived, and Sally dutifully took him out to the barn and directly over to the stall with the nail. "This is the cow over here," she told him.

"What's the nail for?" the man asked.

Sally stopped to think and replied, "I guess it's for hanging up your pants."

● "How many sheep does it take to knit a sweater?" the tourist asked the rancher.

"I didn't even know they could knit," the rancher responded.

● A novice dairyman, having just moved from the city, asked the country agent, "Tell me how long cows should be milked."

The agent looked at the dairyman strangely and replied, "Just like short cows should be milked."

● A city woman visiting a small rural town went looking for a new dress. As she walked past the front window of a boutique, one dress in particular caught her eye.

"I'd like to try on that dress in the front window," she told the sales clerk.

"I'll get it for you," answered the saleswoman, "but we'd rather you try it on in the dressing room."

● An antique collector from the city was passing through a small village. He stopped to watch an old man chopping wood with a wooden axe.

"That's a mighty old axe you have there," he remarked.

"Yup," said the villager, "it once belonged to Sir Walter Raleigh."

"No kidding!" gasped the collector. "It has certainly stood up well."

"Of course," admitted the old man, "it's had three new handles and two new heads."

● Two city slickers, on their way to Alabama, were finding the country music playing on their radio tiresome.

"What do you get when you play country music backwards?" asked one, trying to make light of the situation.

"I have no idea," replied the other.

"You get your girl back, your dog back, your pickup back and you stop drinking!"

● The two travelers knew they were on a no-frill airline when a voice came over the P.A. and announced, "Ladies and gentlemen, the cabin attendants will now begin food service. Please take only one bite before passing the sandwich along."

● A commercial traveler was passing through a small town when he came upon a huge funeral possession.

"Who died?" he asked a nearby local.

"I'm not sure," replied the native, sizing up the stranger, "but my guess is it's the fella in the coffin."

● A city boy was visiting a ranch in Wyoming on his first trip out West. He came to the ranch house with a handful of rattles from an enormous rattlesnake. When one of the ranch hands saw them, he turned pale and said, "Golly, where did you get those things?"

The city boy couldn't understand why the ranch hand was so nervous, but explained, "I just took them off the biggest worm I ever saw!"

● A city chap was crossing a pasture.

"Say, there," he called to the farmer, "is this bull safe?"

Trying to conceal a smile, the man replied, "Well, he's a lot safer than you are!"

● A city fellow inherited his uncle's dairy farm and immediately went to take possession. Shortly after he arrived, a friend visited him to see how things were going. He found the new farmer out in the barn, looking at the lines of dairy cows.

"They sure look great," remarked the friend. "How many head have you got?"

"Don't know," answered the farmer, "they're all facing the wrong way."

● A farmer was in the process of interviewing possible candidates to lease his town property. "Sorry, we don't allow dogs," said the farmer to the obvious city slicker, as he glared down at the tiny Chihuahua.

"Yes, but this is a seeing-eye dog," the potential renter lied desperately. "I'm blind."

"Guide dogs are usually big dogs," growled the farmer skeptically. "I've never heard of a guide Chihuahua."

"What?" exclaimed the prospective renter. "They gave me a Chihuahua?"

● "Papa," asked the city boy vacationing in the country, "what makes roosters crow so loudly every morning?"

"Son," the father said, putting his hand on his son's shoulder, "they're merely making the most of their opportunity before the hens wake up."

COW POKE

● A Harvard Business School-educated economist was on a business trip when his plane was forced to land near a small rural farm town. Settling in for the evening at the town's only motel and obviously disappointed with the accommodations, he called down to the front office and barked, "May I please get turn-down service now?"

The manager, slightly taken aback, took down his name and said she'd take care of the matter shortly. A few minutes later, a gruff old maid appeared at the economist's door and said, "I understand you want turn-down service."

The economist nodded and said, "Absolutely!"

"Fine," the maid growled as she looked him over. "I wouldn't go out with you if you were the last man on earth!"

● Two airline passengers were swapping travel stories.

"I think I must've flown the smallest airline around the last time I traveled," one said. "During an emergency, a little compartment over my seat popped open and a rabbit's foot fell out."

"That's nothin'," challenged the other. "During a flight to Europe once, I was on a plane that was so low budget, they handed out fishing poles at dinner time and flew real low."

● "Is it tough to milk a cow?" asked the city slicker on his first visit to a farm.

"Oh, no," said the farmer. "Any jerk can do it."

● A wealthy woman passing through a small rural community stopped at the post office to buy stamps. "Do I have to stick these on myself?" she asked the clerk.

"No," the clerk replied. "They work a lot better if you stick them on envelopes."

● "You can always tell when a plane is about to hit turbulence," commented Ernest, wriggling in his seat.

"You can? How?" asked the man seated next to him.

"It's when the stewardess serves the coffee."

● An Ohio farmer walked up to a Native American in New Mexico one afternoon.

"How!" he said. "White man hope-um red man feelum heap good today. Ugh!"

The Native American stuck his head inside his tepee and shouted, "Hey, Martha! Come get a load of this guy, will ya?"

● A tourist from the big city was excited to visit a real Indian reservation and see the elders practice some very old traditions. He especially wanted to know how the old Indian men could predict warm and cold weather so accurately. He decided to ask one old wise man, sitting on a tattered rug.

"If I tell you, you must promise never to tell anyone else," the old man told the visitor.

The city slicker leaned close to the old man to hear the secret of his powers.

"Watch white man's wood piles," he said. "If all wood piles big, winter cold."

● "How do you get to St. Louis from here?" called the tourist to the old farmer working in his field along the highway.

The farmer considered the question, then carefully replied, "My brother usually takes me."

● "Did you hear the airline luggage handlers were supposed to go on strike last week, but decided against it?" Stan asked his traveling buddy.

"No," replied Ben. "Why didn't they go through with it?"

"Somebody lost the picket signs."

● Herbert was fed up with his know-it-all cousin from the city. At last, he said, pointing, "I bet you don't even know whether that's a Jersey cow."

"Of course I don't," came his cousin's reply. "I can't see its license."

● Some city boys were hiking in the country. One of them came upon a pile of empty milk bottles.

"Come quick!" he called excitedly to his friends. "I've found a cow's nest!"

● A city boy, visiting his country cousin, was walking through a pasture when he heard a buzzing sound.

The cousin shouted, "It's a rattlesnake! If you go near it, it will strike!"

"Is that right?" asked the city boy. "They've got unions, too?"

● George had recently moved to the country from Chicago. One afternoon he took a trip to the hardware store for a new garden hose. He bought one 30 feet long.

Upon arriving back at his house, George noticed his farmer neighbor standing on his front yard.

Trying to make conversation, George asked his new neighbor, "Just how far do you think a 30-foot hose will go on this lawn?"

The farmer looked strangely at George and answered. "Well, I don't know what they teach you about hoses in the big city, but wouldn't ya say about 30 feet?"

● A ventriloquist was in a tavern in the Ozarks. He thought he could make a couple of bucks if he took his dummy out and told a few jokes about hillbillies. After he told a few, a young feller started to get hot under the collar.

He stood up and said, "I'm gettin' tired of these jokes. Not all of us is dumb ya know."

The ventriloquist said nervously, "I'm sorry, sir. It was all in jest."

The young fella said, "I'm not talkin' to you. I'm talkin' to that smart aleck on your knee!"

● The city man was visiting a ranch for the first time and decided he'd like to ride a horse. But, he soon found himself sprawled in the dust.

"Boy, that horse can sure buck!" he exclaimed, wiping himself off.

"Buck!" snorted the cowboy. "She only coughed!"

● A salesman was passing through a small town in the South, when he happened across a friendly-looking man.

"Say, does this town have any nightlife?" the out-of-towner asked.

"Sure does," the native replied, "but she's ill today."

You May Be A Farmer If...

● You've never thrown away a 5-gallon bucket.

● You have used baling wire to attach a license plate.

● You have used a chain saw to remodel your house.

● You have driven off the road while examining your neighbor's crops.

● You have used a velvetleaf plant as toilet paper.

● You've used your castrating knife to slice an apple.

● You can remember the fertilizer rate, seed population, herbicide rate, and yields on a farm you rented 10 years ago, but cannot recall your wife's birthday.

Chapter 6

The World Outside The Country

"Anytime four New Yorkers get into a cab together without arguing, a bank robbery has just taken place."

—Johnny Carson

● A country couple went into a nondescript little restaurant in the city. Both ordered coffee, but the wife told the waitress, "Please make sure the cup is clean."

"Honey, that's insulting," the farmer whispered to his wife.

"Well," she answered, "the place looks a little unclean."

Moments later, the waitress returned with the coffees and asked, "Now, who gets the clean cup?"

● The farmer was having a tough time lugging his lumpy, oversized garment bag on board the plane. With some help from the flight attendant, he finally managed to shove it in the overhead compartment.

"I don't know how you manage," sighed the attendant.

"Well, I'm never doing that again," snapped the irritated farmer. "Next time I'm riding in that bag, and my son can buy the ticket!"

● Lou and Chuck's sightseeing tour concluded with the most famous New York landmark, the Statue of Liberty. Gazing up at the great statue, Lou said to his traveling buddy, "You know, with the crime rate as high as it is in this city, we may soon see Lady Liberty standing with both hands up."

● Farmer Emil and his wife were window-shopping in New York when a man bolted out of nowhere, practically knocking the couple over.

"That proves it, Franny," Emil said, straightening up, "New York is the only city where you can get run down on the sidewalk—by a pedestrian."

● "Why do they tell you to check in at the gate a whole hour before your plane takes off?" complained Farmer George on his way to the No-Till Conference.

"It's probably for our own good," said Farmer Mel. "The sooner we check in, the sooner we'll find out about the delay."

● Two fun-loving farmers, in town for the weekend, instructed their cab-driver to take them to the nearest casinos.

"Gentlemen, if you want to gamble in this great city, there are three good ways to do it," the cabbie said. "Off-track betting, state lottery or if you're really feeling lucky, ride the subway."

● The little farm boy looked up and down the big city street, then sorrowfully approached a policeman, standing on the corner.

"Sir," he asked, tugging on the man's pant leg, "did you see a lady going by without me?"

● A Nebraska woman attended a cattlemen's convention in the big city. Determined to have something unusual to wear, she made a blouse and embroidered every cattle brand she knew on it.

In the hotel where the cattle folk stayed, she waited while her husband registered, noticing two old cattlemen giving her blouse the once-over.

Finally, one of them remarked in a loud voice, "That critter sure has changed hands a lot, ain't she?"

● Farmer Chuck was getting annoyed at the lack of service in the city restaurant.

"Say, are you the same waiter who took my order?" he asked the man walking briskly over to his table.

"Yes, sir," the man answered.

"Interesting," Chuck responded. "Somehow I expected a much older man."

● A farmer, in the city attending a farm conference, sat down to enjoy a peaceful meal at a corner cafe, eager to relax from his busy day. When the waiter approached him, he told him he'd like a bowl of soup and a kind word.

The waiter brought the soup immediately, but didn't say anything. "Hey," said the farmer, "How about the kind word?"

The waiter leaned over the table and whispered, "Don't eat the soup."

● Two traveling farmers were complaining about the amount of crime running rampant through the nation's cities.

"Geez," commented Farmer Hank, "it's really getting rough out here."

"Know what you mean, pal," Farmer Frank agreed. "Just this morning, some guy stopped me outside the hotel and tried to sell me a watch...with the wrist still in it!"

● The visiting farmer caught the eye of a waiter in the urban restaurant and beckoned him to his table.

"The service here is terrible!" said the irate patron.

"How would you know?" said the cool waiter. "You haven't had any yet."

● An Iowa farmer, on his first visit to the big city, stopped in at a plush cocktail lounge and ordered a beer.

"How much beer do you sell a week?" he asked the bartender.

"About 40 kegs," the man replied.

"I'll tell you how you can sell 80."

"80 kegs?" said the bartender, amazed. "How?"

"Simple," the farmer answered, raising his beer. "Fill up the glasses."

● A cattle baron from the Southwest was visiting a college campus for the first time and took his wife on a tour of the street where all the sororities and fraternities were located.

"Look at that!" he exclaimed, pointing to the neon signs with Greek letters on each house. "What a good idea—everyone has his cattle brand up in lights."

● A farmer, passing a nudist colony, noticed a man with an extremely long beard coming out the front gate.

"Excuse me, sir," said the curious farmer, "but isn't that unusual for a nudist to have such a long beard?"

"Well," came the reply, "somebody has to go for pizza."

● The hog farmer and his wife were celebrating their 20th wedding anniversary at a fancy restaurant in the city. Expecting an extraordinary feast, they were disappointed by a mediocre meal.

"Waiter," complained the farmer, "this meal isn't fit for a pig!"

"Just one second, sir," replied the waiter, "I'll get you something that is."

● A farmer walked into a New York City bank and asked to borrow $5,000. The loan officer asked for some kind of collateral, so the man tossed her the keys to his brand new pickup truck. "It's in your parking garage—just keep it here until I return," he said.

Worried about the bank being responsible for such an expensive vehicle, the banker assigned a security guard to keep a close eye on it.

Two weeks later, the farmer returned and repaid the loan, plus $20 in interest and asked for his pickup.

"Frankly, sir, we're puzzled," the loan officer said, handing the man his keys. “We checked our records and found you are worth millions. Why would you need to borrow $5,000?”

“I didn’t,” the man replied. “But where else in New York City can I park my truck for two weeks for $20?”

● A wise-guy farmer was heading off to Chicago for the annual ag conference. Upon arrival at the airline counter, he said to the clerk, “I’m flying to Chicago. Send the suitcase to New York and the little bag to Cleveland.”

“We can’t do that,” said the clerk indignantly.

“Why not?” asked the farmer sarcastically. “You did it last year.”

● En route to Florida, a farmer and his wife stopped at a gas station to fill up their car. While the farmer pumped the gas, his wife went to use the restroom.

“Here’s the key,” said the attendant. “We keep our bathrooms locked.”

“Why?” asked the woman. “Are you afraid someone will break in and clean them?”

● Farmer Ned was making travel arrangements to attend the annual farming conference.

“You know, hon,” he said to his wife. “I don’t think they should perform random drug testing on all airline employees. Just the ones who keep changing the fares.”

● A farmer’s wife who had just moved into a big city was getting her daughter bundled up for a day of frolicking in the winter weather.

“Now, dear,” the mother advised, “go out and play before the snow gets dirty.”

● A group of farmers had joined up with a farm magazine’s tour abroad. One afternoon, they took a break from touring local farms to do some sightseeing in the city. As they toured the museum, the guide droned on. “And this is where the Magna Carta was signed...”

One of the tourists piped up, “When did this happen?”

The tour guide replied, “1215.”

“Wouldn’t you know?” said the farmer. “We only missed it by 25 minutes!”

● A dusty old cattle rancher checked into a hotel on a trip to the big city. “Excuse me, sir,” he said to the clerk, “but could you please give me a room and a bath?”

The hotel clerk looked at him and replied, “I can give you a room, but I’m afraid that you’ll have to take your own bath.”

● On a trip into town, a rancher stopped at a department store and bought a cigar. He had just started to light up when he was informed smoking was not permitted in the store.

"What?" he exploded. "You sell cigars in here but you prohibit smoking?"

The sales clerk smiled politely and said, "We also sell toilet seats, sir."

● Two farmers were trying to decide which airline to take out of town.

"You know," said one, "the airline I use has made a major improvement in its in-flight food."

"Is that so?" asked the other.

"Yup, on many flights they've stopped servin' it."

RAM BULL

● On a trip into the city, the ranch hand stopped at a fast food restaurant for lunch. Upon ordering, he noticed a sign for a hamburger. It said, "A quarter-pound before cooking."

The inquisitive patron asked an employee, "Well, what is it after cooking?"

After a moment of thought, the employee replied, "I guess it's still a hamburger."

● A farm boy on his first visit to the city had been standing on a busy corner for hours while cars and trucks zoomed by. He couldn't get up enough courage to dart across, but finally spied a man on the other side of the street and called over to him, "Hello! How in the world did you get over there?"

The man across the street cupped his hands and shouted back, "I was born over here!"

● A small town drunk making his first trip to the big city stumbled up to a parking meter and put a dime in it.

As it went up to 60 he exclaimed, "Oh no! I've lost 60 pounds!"

● Russ and Rick, two Arizona horse owners, found an article that caught their eye in one of the fishing magazines. It was about ice fishing in Wisconsin, something they had never experienced, since the idea of a lake freezing over was surely new to them.

"We ought to try it," Russ said.

"Let's go this weekend," replied Rick.

So they loaded everything into the pickup and headed north. When they got to Wisconsin, they pulled into a parking space at a lakeside bait shop to stock up on supplies.

"By the way, you'd better let us have an ice pick," Russ said. They paid for the ice pick and left. In a half an hour they were back, needing two more ice picks.

After about an hour and a half, Russ and Rick were back at the bait shop. "I'll need five ice picks," Rick said.

By the middle of the afternoon, they returned again, asking for more ice picks.

"Say, you guys already bought every ice pick in this place," the bait shop owner said. "What the heck are you doing with them, anyway?"

"Well, they're wearing out so darn fast, and we still don't have our boat in the water," came Rick's disgusted reply.

● A New York cab driver picked up a Kentucky farmer at the airport. "You're from the country, right?" the cabbies inquired.

"Yup," replied the visiting farmer.

"Well, I've got a riddle for you that you can take home with you," the New Yorker said. "I'm thinking of someone who has the same father and mother as I do, but is not my brother or sister. Who can it be?'

The farmer shrugged his shoulders and responded with, "Beats me. Who can it be?"

"Me," the driver said, bellowing with laughter.

The tourist thought for a moment and then laughed. After returning to his farm in Kentucky, he tried the same riddle on his friends at the stables.

After the group tried and tried to answer the mystery riddle to no avail, they gave up and asked who it could be. The farmer flung back his head, slapped his knee and chuckled to absolute silence, "He's a New York cab driver!"

● Farmer Ed and three of his city friends were golfing at a new course one day. Ed was the first to tee off at the first hole. After a few confident practice swings, the farmer "whiffed" his first shot, completely missing the ball.

Surprised, Ed turned to the rest of the foursome and exclaimed, "This sure is one tough golf course!"

● The old cowboy was making his first trip to the big city. At a party he attended, an awe-struck party-goer saw the gawdy ring with an enormous stone on his finger.

"Is that a real genuine diamond?" the guest asked.

"Well, if it ain't," the cowboy drawled, putting his hand on his hip, "I sure been beat out of my $20 and a half!"

● Farmer Hal, who went to the city to see the sights, got a hotel room for the night. Before retiring, he called down to the front desk to inquire about when meals were served.

"We have breakfast from 7 to 11 a.m., lunch from 12 to 3 p.m. and dinner from 6 to 8 p.m.," explained the clerk.

"Look here," inquired the farmer in surprise, "when am I gonna find time to see the town?"

● A retired farmer and his wife took up residence in the big city, for a change of scenery. Feeling like a new woman, the Mrs. decided for the first time in her life to open a savings account in her name.

Filling out the application was a cinch, until she came to the space for "Name of your former bank."

After a slight hesitation, she put down "piggy."

● Two women began chatting on a bus, one of whom was evidently from Chicago and the other from Albuquerque. They entered into a lively discussion of the relative merits of their home towns.

"In Albuquerque," one woman remarked, "we place all our emphasis on breeding."

After a while, the other woman replied, "In Chicago, we think it's a lot of fun, too—but we do manage to catch a show every once in a while."

● A cowboy swaggered into a fancy restaurant and sat down at a candle-lit table. He stared down into a plate of rare steak, just singed slightly, with blood oozing amidst the sizzling juices.

He jumped and stood up abruptly. Then he threw his napkin on the floor and turned to stomp out the door. "Heck," he said, "I've seen cows get well that were hurt worse than this!"

You May Be From A Small Town If...

- **You can name everyone you graduated with.**
- **You get a whiff of manure and think of home.**
- **You know what 4-H is.**
- **You ever went to "headlight parties."**
- **You said a certain four-letter word and your parents knew within the hour.**
- **You ever went cow tipping.**
- **School gets canceled for state sporting events.**
- **You could never buy cigarettes because all the store clerks knew how old you were (and if you were old enough they'd tell your parents anyhow).**
- **Social acceptance in town depended on the approval of five old (rich) ladies that met each morning at the donut shop for the latest dirt.**
- **No place sells gas on Sunday.**
- **You have to drive an hour to buy a pair of socks.**
- **You have ever gone for a walk in the cemetery, on a date.**
- **You don't give directions by street names or house numbers, but you give directions by references ("turn by Armstrongs' Liquor, go two blocks past Andersons', and it's four houses left of the track field").**

Chapter 7

Life On The Farm

"If you work long and hard enough on a farm, you can make a fortune— that is if you strike oil."

—James C. Humes

● Doc Jones was giving the farmer's wife the results of her husband's exam.

"Mrs. Brown," the doctor said, "I'm afraid your husband will never work on the farm again."

"I'll go right in and tell him," Mrs. Brown replied. "It'll cheer him right up!"

● A real estate man was using high pressure tactics to sell lousy farmland.

"Heck, all this land needs is a little water, a cool breeze and some people to farm it," the shifty man said.

"Maybe so," replied the keen farmer, "but that's all hell needs, too."

● "Come quick!" urged the farm wife on the telephone. "Our house is on fire!"

"How do we get there?" asked the fireman.

The woman yelled, "Don't you still have that big red truck?"

● Luke was visiting a chicken farmer. The farmer proudly showed him a very large egg and said, "What do you think about that?"

Luke was impressed. "Well," he said, "I bet it doesn't take very many of them to make a dozen."

● A farmer was holding a pig up to an apple tree to eat. "Wouldn't it save time if you pulled down an apple first, then fed the pig?" asked a passerby.

"Sure," answered the farmer. "But what's time to a pig?"

● A Florida farmer was telling his colleagues at the annual farm convention about the special problems he encountered while farming in his area of the country. "Why, just the other day, I was checking the ponds in my fields and an alligator bit off one of my toes!"

Horrified, one of the other farmers asked, "Which one?"

"Beats the heck out of me," the Floridian replied. "All 'gators look the same to me."

● The health officer called Farmer Tom and told him he could take down the scarlet fever quarantine sign now that all the children had recovered.

"I don't want to take it down," the farmer whined. "Since you put the sign up, we haven't had a bill collector or salesman on the place."

● "Food prices are ridiculous!" the city dweller complained to the farmer. "Never have farm products cost so much."

"You're right," agreed the farmer. "But understand, when a farmer is supposed to know the botanical names of what he's raisin', and the entomological name of the insect that eats it and the pharmaceutical name of the chemical that kills it, somebody's gotta pay for that education."

● "What we're breeding for," explained the experiment station agronomist, "is a dwarf corn with short stalks and smaller ears closer to the ground."

When the speaker paused, a frustrated farmer blurted out, "Shucks, professor, you don't have to breed for that! I been growin' corn like that all my life."

● Two farmers were always trying to outdo each other in agricultural prowess.

One morning, one of the farmers said to his son, "Go over to Smith's and borrow his crosscut saw for me. Tell him I want to cut up a pumpkin."

On returning, the boy said, "Mr. Smith said he can't let you have the saw until this afternoon. He's halfway through a potato."

● A mountaineer's wife came down to visit the local pharmacy.

When the druggist came back to the counter with the woman's prescriptions, she said, "Now be shore and write plain on them bottles which is fer the horse and which is fer my husband. I don't want nothin' to happen to that horse before spring plowin'."

● An old bachelor farmer lived alone on his little farm with one cow. Each evening after milking the cow, he would lift the pail, take a healthy drink of milk, throw the rest over the fence and carefully wash the pail.

Witnessing this ritual, one of his neighbor's asked why. "Well," the happy bachelor said, "chores are done, dinner's over and the dishes are washed."

● A frustrated farmer, on his way to the county fair, had been trying to pass a huge truck for many miles. Every time he tried to go around, the truck driver increased his speed or swerved slightly toward the middle of the road.

Finally, at a stop sign, the farmer pulled alongside the truck driver's window.

"Well?" growled the trucker, glaring viciously.

"Nothing important," was the reply. "I know what you are—I just wanted to see what one looks like."

● "My garden is 100% natural," Helen told her friend. "It has no pesticides, no chemicals and no additives."

"Is that right?" her impressed friend responded.

"However," Helen admitted, "the thing also has no vegetables, fruits or flowers either!"

● Two farmers were bragging about their talents.

"I once made a scarecrow look so natural that it frightened away every single crow that was on my farm," boasted one.

"Heck, that ain't nothin'," replied the other. "I made one that scared 'em so badly that they brought back the corn they stole from me last year!"

● Farmer Cliff trudged toward the breakfast table.

"Look on the bright side, dear," his wife comforted. "In 16 hours, you'll be back in bed."

● A farmer walked into a supply store and asked for 200 chicks for the chicken farm he was about to start. Two weeks later, he returned to the store, requesting another 200 chicks. The owner was curious but didn't say anything. The same thing happened when the farmer returned in yet another two weeks for the same order.

By the time the farmer returned for the fourth time, the store owner's curiosity got the best of him.

"Why on earth do you keep coming back for so many chicks?" the man asked.

The farmer replied, "Well, I guess I must be doing something wrong, but I don't know what. I must either be planting them too deep or too close together."

Taking the store owner's advice that he needed serious help, the farmer sent off a report of what he had done to the local agricultural school, seeking their advice. Three weeks later, the reply came back, stating simply, "Please send soil sample."

- The salesman stopped at a farmhouse and asked the woman if her husband was home.

"He's over there in the cow barn," she said, nodding toward the building.

"I'd like to speak to him. Will I have any trouble finding him?" asked the salesman.

"None at all," the woman replied. "He's the one with the whiskers."

- What do you call a bunch of farmers in a basement?

A whine cellar.

HOG WASH

● The plant pathologist looked down at the group of farmers gathered to hear his speech on soybean cyst nematodes.

"I'm not being paid to speak with you tonight," he announced. "So, that, right off the bat, blows my credibility."

● Cousin Clara was out walkin' through the deep mud after the spring flood when she saw a hat floatin' along the road. She picked it up and much to Clara's surprise, there was a man underneath it.

Clara declared, "Why, ain't it too muddy to be out a walkin' around this here farm?"

"You ain't kiddin'," the man replied. "That's why I'm ridin' my mule."

● Farmer Zeke had been listening to the weather forecast and had not expected the premature weather that had rolled in.

Rather perturbed, Zeke decided to express his feelings in writing: "Dear Weather Bureau: I thought you might be interested in knowing that two feet of partly cloudy just covered my vegetable crop."

● A farmer had some boots made, and they turned out to be too tight. The bootmaker insisted on stretching them for the man.

"Not on your life!" exclaimed the farmer. "Every morning when I get out of bed, I gotta jump right into the field and get to work plowin' and workin' the land. All day long, I watch my crops wither away from lack of rain. Then, I still have to tend to my livestock.

"After supper, I watch the news and hear about the high price of feed and the low price of beef. And all the while my wife is nagging me to move to the city.

"So you see, when I get ready for bed after a day like that and pull off these tight boots, that's the only pleasure I get all day!"

●The meeting of a farm organization had been stormy, and tempers were running hot.

"You, sir," shouted one member at another, "are about the most pig-headed man I have ever met!"

"Order! Order!" exclaimed the chairman. "You gentlemen seem to forget that I am in the room."

● At the funeral of a man who had farmed for years, the priest had many fine things to say. He concluded with the thought that the deceased was probably carrying on his work in heaven.

Two farmers were sitting together. "Good Lord," said one to the other, "don't we ever get to quit?"

● Two farmers had been neighbors for more than 20 years but hardly ever spoke to each other.

One morning, the silence was at last broken. "Hey, Rex! My best cow is ill. She's all bloated. What did you give your cow when she was like that?"

"Liquid paraffin and sulfur," grunted the other.

Two days later, the two men met again.

"Hey there, Rex! I gave my cow liquid paraffin and sulfur like you told me, and she just up and died!"

The other farmer started to walk away as he mumbled, "Mine, too."

● Farmer Dale had just taken his raving mad wife to the asylum for intense treatment.

"What do you think caused it?" his neighbor asked when he returned home.

"Can't imagine!" replied Dale. "I ain't had her off the farm in 40 years."

● ***Poor Farmer Herb, feeling the effects of old age, decided to write his will...***

— I, Herb Miller, leave the following:

— To my wife, my overdrawn bank account. Perhaps you can explain it.

— To my banker, my soul. He has the mortgage on it anyway.

— To my neighbor, my clown suit. He'll need it if he continues to farm the way that he does.

— To the ASCS office, my grain bin. I was planning on letting them take it next year anyway.

— To the county agent, 50 bushels of corn to see if he can hit the market. I never could.

— To the junk man, all my farm equipment. He's had his eye on it for years now.

— To my undertaker, a special request. I find it appropriate for six implement and fertilizer dealers to be my pallbearers; they should be used to carrying me by now.

— To the weatherman, rain and sleet and snow for my day of burial, please. No sense having good weather now.

— To the grave digger, don't bother me. The hole I'm in should be big enough.

Chapter 8

Woe Is Me

"Man invented language to satisfy his deep need to complain."

—Lilly Tomlin

- A farmer won a lottery jackpot of $10 million. "What are you going to do with all the money?" asked a reporter.

The farmer replied, "Oh, I imagine I'll just keep farming 'til it's all gone."

- Trying to sell a farm wife a home freezer, the salesman pointed out, "You can save enough on your groceries to pay for it."

"Yes, I know," the woman replied, "but we're paying for our car on the money we save on bus fare. Then, we're paying for our washing machine on the laundry bills we save and we're paying for the house on the rent we save.

"So, you see, mister," she concluded, "we just can't afford to save any more money right now."

- Two farmers pulled out their brown bag lunches. The first took out a cottage cheese salad. "On a low-calorie diet, Al?" asked his dining buddy.

"No," replied Al, "a low-profit diet."

- Back when gold coins were still being used, a gentleman visiting a farm handed a $5 piece to a child to examine. The child put the shiny coin on his tongue and swallowed it.

The child's father, a man short on his money, scooped up his son and rushed him to the bathroom where he promised he'd make the boy cough up the coin.

The sly farmer returned a while later and put $1.30 in change in the visitor's hand.

"Sorry," the farmer said, shrugging his shoulders. "I guess he digested the rest."

- Two farmers bumped into each other at the local feed market. "Ted, you look kinda pale. What's the matter?" said Jed.

"I think our local savings and loan is in trouble, Ted," replied Jed.

"Aw, why do you say that?" asked Ted.

"Well," explained Ted, "I got their new annual calendar and it only goes through May! Plus, they've changed their name to Westside Savings & Loan and Lawn Mower Repair!"

● "Boy, did I get a great order from Farmer Jones for new farm equipment!" boasted a farm equipment salesman to his competitor. "$200,000!"

His peer looked at him, obviously very doubtful. "Oh, yeah?"

"You don't believe me?" shouted the first salesman. "Let me show you the cancellation!"

● "How was your vacation?" the farmer asked his friend.

"Oh, it was all right," the friend replied, "but have you ever actually felt homesick?"

"Sure," replied the farmer. "Every time I make a mortgage payment on the farmhouse."

● Inflation is bringing us true democracy. For the first time in history, luxuries and necessities are selling for about the same price.

● A recession is a period when you tighten your belt.

In a repression, you have no belt to tighten.

And when you have no pants to hold up, it's a panic.

● Inflation marches on, making it possible for us to live on more expensive farms without even moving.

● After working on Farmer Will's jalopy, the mechanic scratched his head and looked at his impatient customer. "What's the verdict, Bob?" Farmer Will asked sheepishly.

"Well, Will, my advice is to keep that car of yours moving." Bob said.

"That's strange advice. Why should I do that?" questioned Farmer Will.

"Well, if you ever stop," replied Bob, "the cops will think it's an accident."

● Grandpa says, "Running into debt isn't so bad. It's running into creditors that's so embarrassing."

● The farm wife told the butcher, "I'll have a pound of ground chuck, please."

"Lady," replied the butcher, showing his best fresh-ground smile, "with the price of beef being what it is today, we call it Charles now."

● A panhandler stopped Farmer Judd and asked, "Buddy, can you spare a dime for a cup of coffee?"

Judd gave the man a dime and started following him.

"Why are you following me?" asked the annoyed beggar.

Judd replied, "I just wanted to see where you can get a cup of coffee for a dime."

● Grandpa was giving his grandson a bit of his wisdom.

"From birth to 18, you need your parents," he told Little Randy. "From 18 to 30, you need good looks. From 30 to 50, you need a personality. And when you're over 50, that's when you need good investments."

● Farmer Elton filled up his pickup truck with $19.50 of gas and asked the attendant what the station charged on bounced checks.

When the reply was $10, he wrote his check out for $29.50.

● A farmer went to a fortune teller who gazed into her crystal ball and said, "You will be poor and unhappy until you are 45 years old."

"Then what will happen?" asked the farmer hopefully.

"Then you will get used to it."

● In some areas, farming is so bad that a farmer can be brought up on child abuse charges for willing the farm to his son.

● Show me a man with his feet planted firmly on the ground, and I'll show you a man who can't get his pants on.

● The farmer was giving his college-graduating son some advice about the world awaiting him.

"Son," the farmer said, "Just remember this bit of advice: Nothing in the universe travels faster than a bad check."

● A Louisiana farmer, who had a reputation as a skinflint, was visited by a government inspector.

"I hear you are violating the law by paying below the minimum wage," the inspector said.

"Oh, am I?" the farmer cried angrily. "Well, there's Hank who milks the cows and does the chores around the barn. Ask him."

"$200 a week, sir," Hank replied.

"And there's Sammy," the farmer said, calling over the other hired man. "Tell this man your wages."

"$200 a week, sir,"

"And there's the maid, Katie. Ask her."

"$150 a week with room and board, sir," Katie said.

"Any more?" the inspector asked.

"Well, no—only the half-wit," the farmer said. "He gets $100 a week, a bit of tobacco and his food."

"A-ha! Could I speak to him?" the inspector asked.

"Sure," the farmer answered. "You're speaking to him now!"

● "Wise men say that money can't buy happiness," Farmer Ray reminded his forlorn friend.

"Is that right?" his friend replied. "Well, I reckon I'd prefer to find out for myself."

● A little farm girl was sitting under an apple tree with her mother when she asked, "Mama, what happens to old tractors when they stop running?"

"Well, dear," her mother replied, "someone sells them to your father."

● "I helped out a friend in financial trouble," a farmer told a neighbor, "and he was so grateful he said, 'I promise I will never forget you!'"

"So, what happened?" the neighbor asked.

The other man replied, "He was right—he never forgot me. Every time he gets in trouble, he calls me again."

● A collection firm sent an overdue bill to a farmer with an attached note: "This bill is now one year old."

The bill was returned with, "Happy Birthday."

● Old Farmer Roy was struggling with his family's monthly budget.

"You know, Fran," he said to his wife, "we should have saved during the Depression so we could live through prosperity."

● "Running a farm today is like a three-ring circus," Farmer Max complained to his buddy. "While you're trying to balance the accounts, your wife juggles the budget and the banker bounces your checks."

● Two farmers took an afternoon off and went to the race track.

Said one to the other, "I hope I break even today. I sure need the money."

● A farmer brought his new car in for its 6,000-mile inspection. "Is there anything the matter with it?" the service manager asked.

"Well, there's only one part of it that doesn't make a noise," the customer replied, "and that's the horn."

● A genie, eager to help people out during hard times, heard about a farmer who was having trouble making ends meet. So, he appeared on the farm one day and told the farmer he could have one wish.

The farmer responded by saying that he had harvested a bumper crop of corn, but the price was so low, his bills wouldn't get paid. "Tomorrow morning, corn will be $5 a bushel," promised the genie.

A few months later, the genie again visited the farmer and found him still in dire financial straits. "What happened?" asked the genie. "Didn't I raise the price of corn to $5 a bushel the last time I was here?"

"Well, yeah," said the farmer, "but I didn't sell because I thought the market might go higher."

● Two farmers habitually tried to outdo each other with complaints of how tough their lives were. This time, one farmer was complaining about his recent trip south.

"It's nice to be back from vacation. We had awful weather. It rained most of the time," he said.

Of course, the other farmer didn't believe a word he said. "Why, look at you!" he exclaimed. "It couldn't have been too bad. That's a nice tan you have."

The first farmer replied, "Tan, nothin'! That's rust!"

● What can a bird do that seven out of 10 farmers cannot?

Make a small deposit on a car.

● "Says here the average man now lives 32 years longer than the average man did in 1800," said Farmer George, reading from his newspaper.

"He has to," groaned Farmer Will. "How else is he gonna pay off all his bills?"

● "Well," declared Farmer Gilbert at the town watering hole, "I've paid off three cars!"

"Three cars!" exclaimed his buddy. "Wow!"

"Yup," said Gilbert, as he ordered another cold one. "My doctor's, my dentist's and my chemical dealer's."

● Farmer Hank called the automobile dealer after purchasing a new truck. "Was it your company that announced on TV that you recently put a truck together in seven minutes?" he asked the man on the phone.

"Why, yes, sir!" the executive answered proudly.

"Well, then," the disgruntled farmer replied, "I just called to let you know I think I've got the truck."

● "My car is so old," Farmer Earl grumbled, kicking the vehicle's fender.

"It's not that bad," consoled his friend.

"Oh, yeah?" answered Earl. "The insurance policy I took out on that heap covers theft, fire and Indian raids!"

● "Sure we have a strong economy," Farmer Lou told his friend. "The tooth fairy told me so."

● The penny-pinching farmer came home from town one day and said to his wife, "I just went to see our butcher. I said to him, 'Let's stop this nonsense. I want something that's lean, red, tender and doesn't cost more than $1 a pound.'"

His wife asked, "What did he give you?"

The farmer threw a small box on the kitchen table and groaned, "Raspberry Jell-O."

● "Money still talks," the farmer told his wife. "But in this house, it usually says, 'Goodbye.'"

Feeling Overworked?

The following information may shed some light as to why...

● The population of the United States is 237 million. 104 of these people are retired. That leaves 133 million to do the work.

● There are 8 million people in school. That leaves 48 million to do the work.

● Of this 48 million, there are 29 million federal employees. That leaves 19 million to do the work.

● Take into consideration the 14,800,000 who work for state and city governments. That leaves 200,000 to do the work.

● There are 188,000 people in hospitals. That leaves 12,000 to do the work.

● Don't forget the 11,998 people in prison. That leaves just two people to do the work: you and me.

● And you're sitting there reading this joke.

Chapter 9

What Would We Do Without Them?

"Where there is a sea there are pirates."

—*Greek Proverb*

● There was a big field day and farmers came from all over to attend. After a busy day, several people were sitting around a table, discussing what they had seen and learned. As more people arrived, they just pulled up chairs and joined in the conversation.

One of the farmers asked, "Anyone want to hear a joke about a county agent?"

A few of the newcomers looked at each other. One said, "I'm 6 foot and weigh 200 pounds and I'm a county agent. This guy next to me is 6 foot 2 and weighs 220 and he's a county agent. And the guy next to him is 6 foot 4 and weighs 245 and he's a county agent. Are you sure you want to tell this joke, buddy?"

The farmer thought a second and said, "No, I guess not. I don't want to have to explain it three times."

● A secretary from the regional Environmental Protection Agency office picked up the ringing phone. The voice on the other end came from a farmer, who asked, "May I please speak with Agent Jones?"

The secretary replied, "No, sir, I'm sorry. Agent Jones has passed away."

Two minutes later after she hung up, the phone rang again. It was the very same request, and it sounded like the same voice. The secretary patiently explained that Mr. Jones had died.

A few minutes later, the phone rang again. After hearing the same voice again asking for Agent Jones, she asked the caller, "Haven't I told you three times now that Agent Jones has passed away? Why do you keep calling and asking for him?"

There was a short pause and the farmer explained, "I just like hearing you saying it so much."

● "I'm thinking of changing my crop consultant," said Ralph.

"Do what you want," said his friend, "but to me, changing crop consultants is like moving to a different deck chair on the Titanic."

● A machinery dealer wa riding down in the elevator after working late one night when Satan suddenly appeared in the elevator with him. "Don't be afraid," said the devil, "I'm here to help you. I've had my eye on you for quite a while, and I think you're my kind of guy.

In fact," he continued, "I'm prepared to give you more wealth than you can possibly imagine, but I'm going to need your soul. Not only that, but I'm going to need your wife's soul. And your children's souls, and your children's children's souls."

The dealer thought for a minute and asked, "So, what's the catch?"

● A doctor, an ag economist, a little boy and a priest were on a small plane when it began to experience engine failure. As the plane nose-dived down, the pilot wasted no time in pulling on a parachute. As he jumped out of the plane, he instructed the passengers to do the same.

Unfortunately, there were only three parachutes left. The doctor grabbed one and said, "I'm a doctor. I save lives; therefore, I must live."

The ag economist grabbed a parachute and said, "I am an ag economist. I am also the smartest man on earth. I deserve to live!" With that, the economist jumped out of the plane.

The priest looked at the little boy and said, "My son, I have lived a long and full life. You are young and have your whole life ahead of you. Take the last parachute and live in peace."

The little boy handed the parachute back to the priest and said, "Not to worry, Father. The smartest man on earth just took off with my backpack."

● A farm equipment salesman and his wife were at an industry trade show when the salesman saw a customer approaching him. Blanking out, he racked his brain to come up with his name, but remembered that his wife had also met the farmer on several occasions.

"Thank God," he sighed, "I can ask my wife." But before he had time to get the words out, the farmer shook his hand and said hello.

The salesman said to him, "I've got a question for you. What do you call that thing...you know, that thing that's red on one end and has a bunch of prickly thorny things on the sides?"

Thinking the question a bit odd, the farmer replied, "Do you mean a rose?"

The salesman nodded and sighed. Then, he turned to his wife and whispered, "Hey, Rose, do you remember this farmer's name?"

● Research labs have stopped using rats and started using tractor salesmen. Why, you say? Because there are more dealers, the assistants don't get as attached to them and there are some things that rats just won't do!

● The clever country bank teller handed over the bag of money to the robber. Then she told him, "If you'll let me invest that for you, I can get you 10% interest."

● A disgruntled banker called a farmer into his office to review his loans. "We loaned you $1,000,000 to expand your cattle feeding operation," he said.

"Coulda been worse," the farmer replied.

"Then we loaned you yet another $1,000,000 to buy new manure handling equipment and it all broke down," said the banker.

"Coulda been worse," the farmer said again.

"I'm tired of hearing that!" snapped the banker. "How on earth could it have been any worse?"

The farmer grinned, "Coulda been my money."

● As part of his introduction, an ag researcher explained that he was so busy concentrating on his speech that he cut his face shaving.

After the presentation, one farmer offered bluntly, "Next time, concentrate on your face and cut the speech."

● The pork bellies trader, shown a list of current examination questions by his old ag professor, exclaimed, "Why, those are the same questions you asked when I was in school!"

"Yes," said the professor, "we ask the same questions every year."

"But don't you know the students hand them along from one year to the next?"

"Sure," said the professor, unfazed. "But in economics we change the answers."

● When Albert Einstein arrived in heaven, he was told that his room was not yet ready and that he would have to temporarily share quarters with three roommates.

The first introduced himself by saying he had an IQ of 180. Einstein was pleased and assured the man that they would have a splendid time discussing the theory of relativity.

The second roommate was quick to boast about his IQ of 120. Einstein replied, "Great. We can discuss the quantum theory of mechanics and examine some mathematical equations."

The third roommate was sheepish when he revealed his IQ was only 80. Einstein paused, gave him a long look and asked, "So, where do you think corn prices are going this year?"

● The farm machinery salesman was trying to sell a milking machine to an old farmer. "If you're selling these machines way under price, like you say," the suspicious farmer told him, "how can you make a living?"

"Simple," the salesman answered, "we make our money fixing them."

● Clint Howe, an internationally-known expert on growing ultra-narrow row corn, was summoned to testify at a local hearing. While being sworn in, he was asked to state his name and occupation.

"Clint Howe," he replied proudly. "World's greatest agricultural speaker."

Later, his lawyer demanded to know why he felt it necessary to answer that way. Mr. Howe replied, "I was under oath, was I not?"

● A man walked into a country bar with his alligator and asked the bartender, "Do you serve chemical dealers here?"

"Sure do," replied the bartender.

"Good," said the man. "Give me a brew, and I'll have a dealer for my 'gator."

● A univeristy agronomist had just finished giving his talk at a national farm conference when a beautiful woman stopped him in the hall. She was wearing a long fur coat that went from her neck to her toes.

The woman opened her coat slightly to reveal to the pathologist that she was completely naked underneath. "I saw you standing at the podium, giving your speech," she said in a breathy voice. "I think you're wonderful, and I'll do anything you say."

"Well," said the man, "would you like me to elaborate on my ideas

ROAD HOG

from my talk about narrow corn rows or the session on nitrogen utilization?"

- What's black and brown and looks good on a seed dealer? A Doberman.

- What's the difference between a mugger and a farm mechanic?
  A mugger uses a gun.

- A dejected ag salesman entered a telephone booth after losing a major business deal for his partner and him. When he discovered he didn't have the right coin, he called to his partner, "Hey, Steve, lend me a quarter so I can call a friend."
  Steve reached into his pocket and handed his partner two coins. "Here's two quarters, swift. Call all your friends."

- "Here's a dilemma," challenged Farmer Will. "You have a gun with two bullets and you're stuck in a room with a bear, a lion and a county agent. Whom do you shoot?"
  "Couldn't tell ya," answered Farmer Dave.
  "The country agent," laughed Will. "Twice."

- Two farmers stood in line to buy feed at the local farmers' co-op. Leroy remarked to his friend, Jim, "Look, there's a seed dealer, a chemical dealer and a machinery salesman, all riding in a car together. Which one do you think is driving?"
  "Gosh, I dunno," said Jim.
  "A cop!" Leroy blurted out.

- ***Unfortunate Signs The Feed Mill You Deal With Is Going Under***
  — They start paying everyone in sea shells.
  — The company president is now driving a Ford Escort.
  — The company softball team is converted to a chess club.
  — Dr. Kevorkian is hired as an "Outplacement Coordinator."
  — The beer supplied by the company at picnics is in unlabeled cans.
  — The boss casually asks a worker if he knows anything about starting fires.
  — When a worker says, "See you tomorrow," the watchman laughs uncontrollably.
  — The women are suddenly friendly to the dorky personnel manager.
  — The chairman walks by the secretary's desk and says,"Hey! Hey! Easy on the staples!"
  — The annual holiday bash is moved from the Sheraton to the local

Taco Bell.
— The president has a dart board marked with all existing departments in the Company.

● A farm equipment salesman rushed into a tavern and shouted, "A lady just fainted outside! Does anybody have a shot of whiskey?"

The bartender quickly filled a glass and handed it to the man, saying, "It's on the house."

The salesman grabbed the glass, downed its contents and handed it back to the bartender. "Thanks," he said. "That always makes me feel better after I see someone faint."

● A train was approaching the station when suddenly it left the track and zigzagged through the meadow.

After a couple of minutes it returned to the station and the station manger yelled, "What happened over there?"

"Larry, the local crop consultant, was walking on the track."

"That's strictly forbidden!" the manager growled. "I would have run him over!"

"Well, I tried to, sir," said the driver, "but I had to go all the way into the meadow to get him."

● Farmers Giles and Mort were taking a midmorning break when Giles asked, "Hey, Mort, why don't ag economists look out the office window in the morning?"

"I don't know, Giles," Mort replied. "Why not?"

"Because if they did, they wouldn't have anything to do in the afternoon!"

● Driving alongside a rural road and seeing a farmer herding sheep, a man got out of his car and called to the farmer, "I'll bet you $100 against one of your sheep that I can guess the total number of sheep in your herd," he said.

The farmer nodded and to his surprise, the stranger guessed 973, hitting the total number of the farmer's sheep right on the nose, thus allowing him to take one of the farmer's sheep.

After the farmer picked up the animal and began walking back to his car, the farmer called out, "I'll bet you double-or-nothing that I can guess what line of work you're in." The man agreed and the farmer said, "You're an economist from a government agency."

Shocked that the farmer was able to correctly identify his profession, the man turned and said, "You're right! How did you know?"

To which the farmer replied, "Put down my dog and I'll tell you."

Our Four-Legged Friends

"Never send a man to do a horse's job."

—*Mr. Ed*

● A mother, knowing that her young children possessed a dire need to uncover secrets, felt desperate measures were needed to prevent the youngsters from discovering their Easter eggs she had colored in advance.

She took the eggs to the chicken house and put them under an old setting hen.

The rooster came home, saw the eggs, flew into a rage, jumped the fence and beat the living daylights out of the peacock.

● A tenderfoot walked into a saloon and was amazed to see a dog sitting at a table playing poker with three ranchers. "Can that dog really read cards?" he asked.

"Yeah, but he ain't much of a player," growled one of the cowboys. "Whenever he gets a good hand he wags his tail."

● ***Points to ponder...***

— How do they get deer to cross at that yellow sign?
— When a cow laughs, does milk come up its nose?

● There was a big fly buzzing about as a farmer was sitting on his stool milking a cow. The farmer watched the fly circle and land on the cow's ear. In fascination, he watched as it disappeared into the cow's ear.

Then, the farmer looked down to the work at hand and noticed the fly was in the milk pail.

It just goes to prove the old saying that "what goes in one ear comes out the udder!"

● If you think you have influence...just try ordering someone else's dog around.

● As a fisherman ran out of bait, he saw a snake with a frog in its mouth. Thinking creatively, the man poured some whiskey near the snake and watched as the snake dropped the frog to gulp up the liquor. The fisherman then scooped up the frog that had been let loose for bait.

A few minutes later, fishing from the bank, the man felt a tap on his leg. It was the snake with another frog.

● The next time you call your dog a dumb animal, just remember who he has working to support him.

● A man walked into a bar and found a horse serving drinks. "Bet you're surprised to see me here," said the horse.

"I'll say," said the man. "The cows must have sold the place."

● A poultry farmer was out in his coop, tending to his hens, when much to his amazement, the hens pulled out little signs posted on sticks and started to walk around in a circle.

"What the heck?" exclaimed the astonished farmer, rubbing his eyes.

"We refuse to lay any more eggs," the largest hen reported sternly. "The sisters and me are tired of working for chicken feed!"

● A group of farm kids were playing a softball game, when a home run sailed over the barn and landed in the hen yard.

Looking it over, a rooster said, "I don't want to complain, girls, but look what the neighbors are doing."

● An old-time prospector listened each morning for his pack animal to bray. When the donkey brayed once, he knew it would be clear and a good day for seeking gold. When the animal brayed twice, he knew the weather would be inclement, a day to stay in camp.

The donkey was his weather burro.

● A dog limped his way into the local saloon and sat down. The bartender asked, "What can I get you?"

"Nothing," the dog replied with a snarl. "I am just looking for the man who shot my paw."

● A farmer walked into his backyard one morning to find a gorilla in one of his trees. He called a gorilla-removal service, and soon the serviceman arrived, armed with a stick, a pair of handcuffs, a shotgun and a Chihuahua.

"Now listen carefully," he told the farmer. "I'm gonna climb this here tree and poke the gorilla with this stick until he falls to the ground.

"The trained Chihuahua will then go right for his, uh, sensitive area, and when the gorilla instinctively crosses his hands in front to protect himself, you slap on the cuffs."

"Got it," the farmer replied. "But answer me this, what's the shotgun for?"

"If I fall out of the tree before the gorilla," instructed the man, "shoot the Chihuahua."

● A lieutenant and a monkey were sent up together in a space capsule. Each had a phone beside his chair.

After a few minutes, the monkey's phone rang. The monkey listened to a few instructions and carried them out carefully. The same thing happened about every 10 minutes, and the lieutenant began to worry because his phone didn't ring even once.

Finally, after about six hours, it rang and the relieved astronaut lunged for the phone, answering in his most professional and military-sounding voice. The message he heard was short and precise: "Lieutenant, please feed the monkey."

● A panda walked into the local diner, sat down and ordered a sandwich. After he ate the sandwich, the panda pulled out a gun and shot the waiter. As he started to leave, the manager shouted, "Where are you going? You just shot my waiter and you didn't pay for your sandwich!"

The panda yelled back at the manager, "Hey, man, I'm a panda. Look it up!"

The manger opened his dictionary and found the following definition for panda: "A bear-like animal of Asian origin, characterized by distinctive black and white coloring. Eats shoots and leaves."

● A farmer noticed another man and a dog sitting in front of him in a movie theater. To his surprise, the dog laughed in all the right places. After the movie, the man expressed his shock to the owner.

"I was surprised, too," admitted the owner. "He hated the book."

● A minister sold a horse to a fellow and told him the critter was trained to go when the rider said, "Praise the Lord," and to stop when the rider said, "Amen."

The new owner mounted the beast, said, "Praise the Lord" and the horse raced away. Becoming excited, the rider kept saying, "Whoa," with no effect on the animal. Then he remembered and said, "Amen."

And the horse stopped abruptly. The rider looked down and found the animal had stopped at the edge of a gigantic cliff. Wiping his brow, he sighed, "Praise the Lord."

● A devout farmer went backpacking one summer in the Rockies. He was hiking along a mountain trail when he heard a crashing behind him—

a ferocious grizzly bear!

The religious farmer started to run along the trail, praying fervently. He burst out of the woods and the path widened. He thought he might be home free until, suddenly, he found himself facing a steep rock wall.

With the growling bear quickly approaching, the farmer prayed, "God, if you've ever answered my prayers, do so now. Please let this beast get religion!"

Sure enough, the farmer turned around to see that bear standing quietly, his paws folded and his furry head bowed. But then he heard the bear reciting, "Bless us, oh Lord, and these thy gifts which we are about to receive..."

● What did the termite say when he walked into the saloon?
Is the bar tender here?

● A new farmer came back from town one afternoon to find his dog with his neighbor's pet rabbit in his mouth. It was obvious to the farmer that the rabbit had been killed and he started to panic. Being new to the area, the farmer didn't want his relationship with his neighbor starting off on the wrong foot.

The farmer decided to try to cover up the whole thing. The quick-thinking farmer took the dirty, chewed up rabbit into the house and gave it a bath, blow-dried its fur and put the rabbit back into the cage at the neighbor's house, hoping the guy would think it died of natural causes.

A few days later, the neighbor leaned over his fence and asked the farmer, "Did you hear that Fluffy died?"

The nervous farmer stumbled around and said, "Um...no...what happened?"

The neighbor replied, "We just found her dead in her cage yesterday."

"I'm sorry to hear that," said the farmer.

The neighbor added, "But the weird thing is, she died last week."

● If you think you have problems, how would you like to be a bee with hay fever?

● The teller selling $5 tickets at the race track was astonished to see a horse step up to the window and ask to bet on himself.

"What's the matter?" snorted the horse. "Are you surprised that I can talk?"

"Not at all," said the man. "I'm surprised that you think you can win!"

● Earl had spent the entire afternoon drinking grasshoppers. Finally, he staggered from the bar and fell onto a lawn, only to come face to face with a real live grasshopper.

"Hi," Earl said, "I've spent all afternoon downing drinks named after you."

"Really?" replied the grasshopper. "You mean there's a drink named Albert?"

● Three turtles decided to have a cup of coffee. Just as they went into the cafe, it started to rain. The biggest turtle said to the smallest one, "Go home and get the umbrella."

So the little one said, "I will if you promise you won't drink my coffee."

"We won't," promised the other two.

Two years later, the big turtle said to the middle turtle, "Well, I guess he isn't coming back, so we might as well drink his coffee."

Just then, a little voice called from just outside the door: "If you do, I won't go."

● Two buxom hens were pecking away in the barnyard. Suddenly, one of them looked over her shoulder and said to the other, "We'd better separate. Here comes that cross-eyed rooster, and we don't want him to miss both of us!"

● Two ants were racing at great speed across a cracker box left behind at a picnic.

"Why are we running so fast?" the first ant asked.

"Don't you see?" the second replied. "It says right here, 'tear along the dotted line.'"

● What's the difference between a coyote in Wyoming and a flea on a dog?

One howls on the prairie; the other prowls on the hairy.

● A poultry farmer was having difficulty with his hens not laying. He had a parrot who could be taught to say anything, so he taught him to say, "Let's go, girls; let's go" and put him in the henhouse that night.

The next morning, the farmer went eagerly to check out the results. He found the parrot high on the ceiling, with nearly all his feathers plucked off. Below was a very angry rooster.

"Wait a minute, wait a minute now," screeched the bird, "I'm only here in an advisory capacity!"

● The farm boss stormed up to one of the hired hands and hollered, "Dunker, I heard you went to the baseball game yesterday instead of coming to work!"

"That's a lie, boss," Dunker said, pulling something from a bag, "and here's the fish to prove it."

● "Why don't elephants go to college?" Bradley asked his school buddy.
"I don't know; why not?" the friend answered.
Bradley answered, "Because few of them finish high school, silly!"

● A farmer brought his sick parrot to the vet. "What seems to be wrong?" the doctor asked him.

The farmer started to tell the vet, when the parrot interrupted. "Hold on, buster," he screeched. "I'm not like your stupid cat. I can talk for myself."

● A Farmer's Guide to Telling If Your Cow has Mad Cow Disease

— She refuses to let you milk her, saying, "Not on the first date."
— Your cow appears on Oprah, claiming to be a horse trapped in a cow's body.
— You find your cow hiding secret plans to burn down half of Chicago.
— Your cow quits the dairy business and applies for a job at Burger King.
— Your cow gets a silicon implant for her udder.
— She keeps wanting to chew other cows' cuds.
— Your cow seems to actually enjoy being hogtied.
— She insists that Milk Duds are the results of stupid cows.
— Your cow asks you to brand her again, but only if you'll wear something risque this time.

● A cowboy fell off his horse and broke his leg way out in the middle of nowhere. The trusty animal grabbed his master's belt in his teeth and carried him to a safe place under a tree. Then he went looking for a doctor.

After hearing the miraculous story, a friend praised the horse's intelligence.

"Hell, what's so amazing about him anyway?" the cowboy griped. "He came back with a veterinarian."

● The farm hand returned to the barn one night to pick up his lunch box. No one was in the farm office except the farmer's big dog, emptying the garbage. The salesman thought he was hallucinating until the dog looked up and said, "Don't be surprised, buddy. This is just part of my job."

"Unbelievable," the man muttered. "I can't believe my eyes. I can't wait to tell the boss what a prize he has in you. An animal that can talk!"

The dog pleaded, "Please...no..don't. If that bum finds out I can talk, he'll have me answering phones."

Lookin' For Love

Chapter 11

"I want a man who's kind and understanding. Is that too much to ask of a millionaire?"

—Zsa Zsa Gabor

● The old maid was rocking on her front porch with her old Tom cat in her lap when suddenly, from a puff of smoke, a fairy appeared.

"You have been good all your life," said the fairy. "I've been sent to grant you three wishes."

"I don't believe it," replied the maiden lady, "but to prove it, turn my rocking chair into solid gold."

No sooner had she spoken than the rocker turned to solid gold.

"What's your next wish?" asked the fairy.

"I wish I was a young and beautiful girl."

Sure enough, she became a young and beautiful maiden.

"And your last wish?"

"I wish my old Tom cat here was a handsome young man."

You guessed it. The cat became a handsome young man who was sitting on her lap. The young man threw his arms around her and began to kiss and hug her. Then he said, "I'll bet your next wish would've been that you hadn't sent me to the vet last spring."

● The boyfriend was studying the check at a posh restaurant in town when his girlfriend noticed his face turning pale.

"Danny, you look ill," she said. "Is it something I ate?"

● A city gal ordered a filet mignon, the most expensive dish on the menu. The waiter looked at her escort and said, "And what do you wish, sir?"

"I wish I hadn't brought her," moaned the farmer.

● A pretty flight attendant had her hands full fending off two persistent men who obviously had had too much to drink. The man in the front of the plane was doing his best to persuade her to come to his apartment. At the back of the plane, the other was trying for an invitation to her apartment.

As the plane headed for the runway, the front-seat pest handed her a key and slip of paper on which he had written his address. "Here's the key to my place and the address," he whispered. "See you tonight?"

"OK," she said, smiling sweetly as she headed for the drunk at the rear. The clever attendant handed him the key and the slip of paper and said, "Now, don't be late."

● Dear Bill: Can you ever find it in your heart to forgive me? I know now that you were the only man I have ever loved. It's taken me some time to figure things out, but I finally have my life together, and I want you to be a part of it.

Please call me soon. I would love to see you again. I will always love you!

—Jenny

P.S. Congratulations on winning the lottery.

● A farmer was dating a bored rich girl from the big city, who seemed to have been everywhere and seen everything. He was stumped, trying to find something that would impress her. Finally, he came up with an idea. "I have a friend that just got a job working for NASA. He thinks he can get us to the moon!"

The girl grimaced. "The moon? I'm not going there—I've heard it has no atmosphere."

● Two young boys were talking as they played on the swing set. "This is such a weird age to be, don't you think?" said one.

"How's so?" asked the other.

The boy looked sad. "Well, I don't know whether to give a girl my seat on the bus—or race her for it."

● My sister is dating an x-ray technician. I don't know what he sees in her.

● "I'm a real animal lover," said the vegetarian, introducing herself to her blind date, Buck.

"So am I," agreed Buck, "especially with some potatoes and gravy."

● A big game hunter was showing off his bear skin rug to his date. "It was either him or me," he bragged.

"Good choice," said the woman. "He makes a much better rug."

● The drugstore cowboy from the big city, whiling away his time in the country, attended his first country auction. The auctioneer, assisted by his lovely daughter, went into his spiel. The girl smiled and nodded to the city boy. He smiled and nodded. This went on for the duration of the auction.

At the end, the blushing city boy was shocked to learn he had bought two dozen duck eggs, a pair of mules, a churn, an electric milking machine and a 40-acre farm.

● In the town's public library, the feed salesman with his new card approached the pretty librarian.

"Do you mean to say," he asked, "that with this card I can take out any book I want?"

"That's right, sir," she answered.

"And may I take out audio cassettes, too?"

"Yes, you may."

"And may I take you out," he boldly asked.

Drawing herself up to full height, the woman replied, "No, the librarians do not circulate."

● Leonard, a boy of 13, was puzzled over his girl problems and discussed it with his friend, Sam.

"I've walked with her three times," he told Sam, "and carried her books. I bought her an ice cream soda twice. Now, do you think I oughta kiss her?"

"Naw, you don't need to," Sam decided after a moment of thought. "You've done enough for that gal already."

● A farmer walked into the local bar after a long day at work.

"What's wrong with you, Larry?" asked the bartender. "You look awful."

Larry sat there a moment staring into his beer. Finally he said, "I asked this woman today if she thought she could learn to love someone like me."

"Well, that's good, Larry! You need to get out there and meet some women!" Larry's friend encouraged. "But, tell me. What's the problem?"

Larry looked the man behind the bar squarely in the eyes and said, "Her reply was 'How much like you?'"

● The farm girl and her date were sitting closely together on the couch in her basement. Suddenly, she looked up at the clock on the wall. "Sometimes my father takes things apart to see why they don't go," she said.

"So what?" replied her date.

"So, you'd better go!" insisted the girl.

● Hank was about to go downstairs to see his daughter's boyfriend out of the house. His wife tried to calm him down. "Now, Hank, don't forget how we were when we were young!"

"That does it," Hank said. "Out he goes!"

● A stunning blond waitress at the local restaurant was constantly being asked by the male patrons for her phone number. However bothersome all these requests seemed to be, the little lady always obliged them with a smile.

The moment of disillusionment came, however, when they called the number and a voice answered: "Pest Control Service."

● A young lady visited a computer dating service and listed her requirements. On her application, she noted that she wanted someone who liked people, wasn't too tall, preferred formal attire and enjoyed water sports.

The computer followed her wishes to the letter. It sent her a penguin.

● The young man was very shy, and after his girlfriend had flung her arms around him and kissed him because he had brought her a bouquet of flowers, he got up and started to leave.

"Oh, dear," the girlfriend said. "I'm sorry if I offended you."

"Oh, I'm not offended," he blushed. "I'm just going out for more flowers."

● A boy becomes a man when he'd rather steal a kiss than second base.

● A girl and a very handsome farm boy were walking along a road together. The farm boy was carrying a large pail on his back, holding a chicken in one hand and a cane in the other, leading a goat.

"I'm afraid to walk here with you," the girl said. "You just might try to kiss me."

The farm boy answered, "How could I, with all these things I'm carrying?"

"Well," the girl said with a twinkle in her eye, "you might stick the cane in the ground, tie the goat to it and put the chicken under the pail."

● "Young man," said the girl's father, "we turn out the lights at midnight in this house."

"Gee," said the boyfriend as a big smile overtook his face. "That's darn nice of you, sir."

● "Get my bag! A man just called and said he can't live another minute without me!" the doctor yelled, rushing out of his study.

"Relax, Dad," the daughter said. "I'm sure that call was for me."

● A drive-in theater is a place where a young man turns off the ignition and tries out his clutch.

● The first date was not going as well as it should have.

"I hope my cigar won't bother you," the man said, lighting up.

"Not at all," the woman replied, "if my throwing up won't bother you."

● "You've been out with worse-looking fellows than me haven't you?" Patrick asked Lisa.

When she failed to reply, he declared, "I said, you've been out with a worse-looking fellow than I am, haven't you?"

Lisa finally replied, "I heard you the first time. I was trying to think."

● Old lovers never die—they just cruise along in front of you on the freeway.

● A farm hand was showing cattle at the county fair. As he wandered around the grounds one afternoon, he spotted a cute girl. After a few minutes of gazing at her from afar, he gathered up the guts to go talk to her.

"Say, haven't I seen your face somewhere before?" came the awkward attempt at contact.

The girl, obviously uninterested, looked at him strangely and replied, "Uh, I don't think so—it's always been right here between my two ears!"

● It had been a year since Widow Smith's husband, a cash crop farmer, had died. It had been a long, lonely year, the widow admitted, as she attended a church service.

The minister was very kind in remembering the widow. He suggested at the service, "In honor of Widow's Smith's presence here today, we will have her choose the first three hymns."

A smile crossed the widow's face as she rose and pointed out three good-looking men. "I choose him and him and him!" she exclaimed.

● Denny and Rita were out on their first date when Denny expressed his version of the ideal woman. "I'm looking for a wife who likes to cook, sew, clean house and who doesn't go out with her girlfriends, drink or smoke."

"Well," snickered Rita, "why don't you go over to the cemetery and dig yourself up one?"

● The young farmer and his date, both under the same umbrella, were walking in the rain on their way to the movie theater.

She looked up at him demurely. "You do realize," she breathed, "that you wouldn't have to go out in this kind of weather if we were married."

● Little Ralphie faced his sister's boyfriend and demanded, "Why do you come to see my sister all the time? Don't you have one of your own?"

● The pretty young girl had just broken off her courtship with a young doctor.

"Do you mean to tell me," exclaimed her girlfriend, "that he actually asked you to return all the presents he had given you while you were together?"

"Not only that," she replied, "the rat also sent me a bill for 44 house-calls!"

● The young farmer looked at the high prices on the night club menu. He turned to his date and said, "What will you have, my lovely, plump little doll?"

● Two sisters had lived alone in their small farmhouse for many years. They carefully guarded their cat Tilly, never letting her out of the house for fear of what might befall her.

One of the sisters finally married and went on her honeymoon. A few days after leaving, the other sister received the following telegram from the bride: "Take my advice—let Tilly out."

Bald Is Beautiful

"Do you think women will ever be the equals of men?" the young woman asked her mother.

"Maybe, dear," replied the mother, "but only until we are able to sport a large bald spot on our heads and still think we are handsome."

Chapter 12

The Newly-Wed

"All forms of gambling are frowned upon by the church—except marriage."

● When a not-too-handsome young farmer proposed marriage, he said, "I know I ain't much to look at."

The young lady replied, "That's all right. You'll be out in the field most of the time anyway."

● The new bride stopped at the druggist's for a refill on sleeping pills.

"I don't know what I would do without them," she sighed. "I'd never get any rest."

"Be sure not to take too many," cautioned the druggist.

"Me?" said the bride in surprise. "Oh, I don't take these. I give them to my husband."

● A young farm mechanic and his bride were at the altar. When the minister asked the bride if she took the groom for richer or poorer, she piped up confidently, "For richer!"

● I went to a wedding this weekend, but I don't think the marriage will last. When the groom said, "I do," the wife said, "Don't use that tone with me!"

● When the young man asked the father if he could marry his daughter, the farmer immediately said yes.

"But, Daddy, how can I leave Mother?" his daughter whined.

"No problem," the farmer said. "Take her with you."

● A young newlywed woman came home from work and found her husband upset. "I feel awful," he said. "I was ironing your favorite suit and I burned a big hole right through your skirt."

"Don't worry about it," consoled the wife. "Remember, I bought an extra skirt for that suit."

"Yes, and it's a good thing that you did," the groom answered. "I used it to patch the hole."

● A rural minister placed an advertisement in the town paper for a handyman. The next morning a well-dressed young farmer came to the door. "Can you start a fire and have breakfast ready by 7 a.m.?" the minister asked him.

The farmer said he thought he could do that.

"Can you polish silver, wash the dishes and keep the house picked up and the lawn mowed?"

"Look," said the young man, "I came to make arrangements for my wedding, but if it's going to be anything like that, I think I'd rather forget the whole thing."

● Shotgun wedding: A case of wife or death.

● A very chic young woman walked into a furniture store and sought out one of its decorators. She wanted advice on how to augment her present furnishings.

"What," asked the decorator, "is the motif: modern, oriental, provincial, early American?"

"Well," was the woman's frank reply, "we were married only recently, so the style of our home is sort of early Matrimony—some of his mother's, some of my mother's."

● A cheery young farm couple were at the country store when the wife happened upon some insect poison at the check-out. "I think I might need to get some of this," the young wife thought aloud.

"What for?"asked the husband.

"To put in your dinner tonight," the wife teased.

"Couldn't hurt anything," the husband said with a smirk.

● "Now, dear, what'll I get if I cook a dinner like that for you every day of the year?" the proud young bride asked her husband.

"My life insurance," came the sarcastic reply.

● The mountaineer called on the doctor for some help.

"Doc, I want you to look at my son-in-law," he said. "I shot him yesterday and nicked him in the ear."

"Shame on you! Shooting at your son-in-law," the doctor scolded.

"Well, he wasn't my son-in-law yet when I shot him."

● The honeymoon is really over when he phones to say he'll be late for dinner, and she's already left a note saying it's in the refrigerator.

● "My dear," the new husband said sweetly, "there's something wrong with this cake. It doesn't taste right."

"That's just your imagination," the bride reasoned. "The cookbook says it's delicious."

● The couple was shopping for wedding rings.

"I don't want too wide or too tight a band," the young man said. "It might cut off my circulation."

"It's going to do that anyway," the girl said, smiling meaningfully.

● If a bride always wears white as a symbol of purity, why does the groom usually wear black?

● The new bride ran to the older lady across the street, greatly upset because her husband had gone hunting.

"Don't worry, dear," counseled the experienced matron, "he'll come back safe."

"Oh, it's not that," said the worried bride. "He's gone to shoot craps and I don't know how to cook 'em!"

● The groom was bragging about what a thoughtful wife his new bride was. "When I go home at night, everything is ready for me," he said. "My cigar and slippers are laid out, the evening paper is beside my easy chair, dinner's on the stove—and always plenty of hot water."

"What's the hot water for?" his friend asked.

The groom looked at his buddy oddly and said, "You don't think she'd expect me do the dishes in cold water, do you?"

● The young farm hand finally got up the nerve to talk to his best girl's father. "Are you afraid I'm too young to marry your daughter?" he asked nervously.

"Not at all," the father replied. "You'll age fast enough."

To Honor And Behold

"And so they got married... and tried to live happily ever after anyway."

● Two woman were overheard talking at the town's class reunion. One said to the other, "Did you ever look at at guy and wish you were single again?"

"Yes," came the reply.

"Really? Who was it?" asked the first.

The other one said, "My husband."

● Little Noah asked his father one day, "Daddy, how much does it cost to get married?"

"I'm not sure, son," Noah's father answered. "I'm still paying for it."

● The chief effect of love is to drive a man half-crazy; the chief effect of marriage is to finish the job completely.

● Young Jack posed a question to his fatner. "Dad, I heard that in some parts of Africa a man doesn't know his wife until he marries her. Is that true?"

"Son," the father replied, "that happens in most countries."

● A recently engaged farm hand was receiving marriage advice from his boss.

"Let me offer up this bit of wisdom, son," he said. "Married life can be very frustrating. In the first year of marriage, the man speaks and the woman listens. In the second year, the woman speaks and the man listens. Now in the third year, they both speak and the neighbors listen."

● As a special treat for his wife's birthday, Ted baked a cake all by himself. However, when they returned home for cake after dinner at the local restaurant, Ted was upset to learn that his dog had eaten the cake.

"Don't worry," his wife comforted him. "I'll buy you another dog."

● Two farm wives were talking at the supermarket. "Did you buy a new dress, Helen?"

"Yes," Helen replied, "I bought it for my birthday. Harvey can never remember when my birthday is, so he just lets me get something for myself."

Ethel said, "It must drive you crazy to have a husband who can't even remember your birthday."

"Are you kidding?" Helen responded with a big grin. "This is my third one this year."

● Two farm wives were shopping and witnessed a man opening the car door for his wife. "Isn't it nice that chivalry still exists in some women's lives!" one of the wives commented.

"I'll say this though," the other responded cynically, "either the car is new or the wife is."

● Two aging farmers were exchanging their views on marriage. "You mean to tell me that you've been married 25 years," said Judd, "and your wife still looks like a newlywed?"

"No," answered Budd, "I said I've been married 25 years and she still cooks like a newlywed!"

● Kids always know when there are guests downstairs—they hear Mom laughing at Dad's jokes.

● Farmer Willy was talking to his friend about his wife's cooking. "I wouldn't say Laura is a bad cook," the man said, "but when she takes the plastic off the microwaved dinner, she's throwing away the best part."

● The hardest thing for most wives to get used to after marriage is being whistled for instead of being whistled at.

● A young farmer was proudly admiring his brand-new tractor. It had cost him a whole year's income.

Noting a bit of dust on it, he called to his wife. "Honey, do you have an old rag?"

"Why, yes, dear," she replied sweetly, "but I'm wearing it."

● This conversation was heard at a small-town diner. "My wife used to make the most delicious roast beef, vegetable dishes, salads and deserts," the man complained to the waitress. "Now all she ever makes are reservations."

The farmer quickly chimed in, "She must have given my wife her recipe."

● One farmer whose wife kept telling him how to drive finally said, "I wish you'd stop your back-seat driving."

The wife thought about it for a minute and said, "OK. I'll make a deal with you. I'll stop driving from the backseat if you'll stop cooking from the dining room table."

COLD TURKEY

● Farmer Arthur thoughtfully bought his wife an electric blanket, since she always complained about being cold at night. She was a little reluctant about sleeping under all those wires, but her husband assured her it was perfectly safe. Within a few minutes, Arthur's wife was dozing off contentedly.

What Arthur didn't know was that his wife had put a turkey in the oven to bake all night on low heat. When Arthur awakened in the middle of the night and smelled something cooking, he reached over and shook his wife, "Marge, Marge, honey!" he cried. "Are you all right?"

● When a man puts his wife up on a pedestal, it's usually so she can reach the ceiling with the paint roller.

● "Doctor, my husband has laryngitis," said the farm wife.

"Well, it's a virus. There's not really anything I can do to cure it," the doctor said.

"Cure it?" replied the woman. "I want to prolong it!"

● A rancher's wife proudly modeled her new fur coat at a posh restaurant on a trip to the city.

"That's really a beautiful fur coat," a friend remarked. "But don't you pity the poor beast who suffered so that you might have it?"

The woman looked at her, surprised. "Why are you suddenly concerned about my husband?"

● Husband: A man who wishes he had as much fun when he was out as his wife thinks he does.

● "You will soon meet a handsome rich man who will shower you with gifts and propose marriage," the fortune teller said.

"Before you continue with this wonderful fortune," the woman interrupted, "what happens to my husband and six kids?"

● A farm wife touring Greece rented a car and drove out to one of the ancient temples crumbling under the centuries. Posing near a huge fallen column, she asked a fellow tourist to take a snapshot.

"But don't get the car in the picture," she warned. "My husband will think I knocked this place down!"

● *How to tell if your computer is fit to be a husband:* It only does what you tell it to and it keeps cutting you off just when you think you finally made a connection.

How to tell if your computer is fit to be a wife: No one but its creator understands its internal logic and you find yourself spending your paychecks on accessories.

● A prim old lady was given her first martini. She sipped it, then said, puzzled, "How odd. It tastes just like the medicine my husband has been taking for the last 30 years!"

● Farmer Herb, visiting an art museum with his wife, stood for minutes rapturously looking at a painting of a woman dressed only in a few leaves.

Finally, his wife snapped at him, "What are you waiting for, Herbert? Autumn?"

● "I believe my husband is the most generous man on earth," exclaimed the proud wife.

"How's that?"

"I gave him six of the loveliest ties for Christmas and he took them right down and gave them to the Salvation Army."

● "Don't judge your wife too harshly for her weaknesses," Farmer Pete warned his son. "If she didn't have them, chances are she never would have married you."

● Alfred and Ray, a newlywed, were talking over a game of poker.

"Heck, I never thought you'd get married, Ray!" Al said. "Just what made you do it?"

"I don't know," replied Ray, "I guess I just got tired of going to laundromats, eating in restaurants and wearing socks with holes in 'em."

"Funny," laughed Al, "those are the same reasons why I got divorced!"

● Farmer Dwight decided to be kind to his wife. He brought home some flowers and candy, kissed her at the door and said, "Come on, honey, slip on your best dress and we'll go out to dinner and a movie."

Dwight's wife broke into tears, "It was bad enough today when the baby fell down the stairs, and I broke my best platter and burned my hand...and now you come home drunk!"

● "It says here in this magazine that most accidents occur in the kitchen," remarked the wife.

"I know," grumbled her husband, "and I have to eat 'em and pretend I like 'em."

● Farmer Jeff came in one evening to find the whole house a mess.
"What on earth happened?" he asked his wife.
"You're always wondering what I do all day," she replied. "Well, now you know. I didn't do it today!"

● Chatting with another farmer at the co-op annual meeting, Benson confided, "My wife swears that if I die, she will forever remain a widow."
"Hmm," commented his companion, "I suppose she thinks there isn't another man like you."
"No," replied Benson grimly, "she's afraid there is."

● Every light went out at the Hawkins farm and Pa and Ma went down to the cellar to investigate.
"Put your hand on that wire, Ma," Pa told his wife, "and tell me if you feel anything."
"Nothing at all," reported Ma.
"Good," said Pa, "now just don't touch the other one or ye'll drop dead."

● After the farmer's frail and elderly wife had broken her leg, the doctor put it in a cast and warned her not to walk up or down stairs. After a month of healing, the doctor removed the cast.
"Can I climb stairs now?" she asked.
"Yes," the doctor said.
"Thank God," she exclaimed, "I was getting so sick of climbing up and down the drainpipe."

● "Is your wife as pretty as she was the day you married?" one farmer asked the other.
"Yes," he replied after some thought, "but it takes her longer now."

● "Honey," the farmer said to his wife, "I would be the happiest man in the world if you could bake bread like my mother used to bake."
"Dear," she replied, "I would be the happiest woman alive if you could make dough like my dad used to."

● You can say this about marriage—it teaches you loyalty, forbearance, self-restraint, meekness and many other qualities you wouldn't need if you were single.

● "Boy, I sure do wish we had my mama's recipe for her chicken with gravy dish," commented the husband as he poked around his plate.

"What do you mean?" asked the wife. "I've had that recipe for years! Why, the meal you're eating right now was made from that very recipe."

"Then what do you add?" asked the husband, "to make it not taste so good?"

- A deceased bachelor left all his properties to three ladies to whom he had proposed marriage—and who had refused.

"It is because of their refusals," he said in the will, "that I owe all my earthly happiness."

- Don't marry for money. You can borrow it cheaper.

- Marriages are made in heaven, which is probably why most married couples are always harping on each other.

- "Do you know what today is?" asked the wife of her husband, who was hurrying off to work.

Hesitating only a moment, and flashing a smile, he replied, "Sure, I remember."

The husband was determined not to be caught again forgetting their anniversary, so that night, he returned home with candy, flowers and some jewelry.

She was overjoyed...as he said smugly, "You see, I did remember what this day is."

To which she warmly replied, "Indeed you did, darling, and you've made it the most wonderful and happiest groundhog day of my life!"

- Two farmers were playing golf when a funeral procession went by. One farmer stopped, took off his hat and watched.

"That's a nice gesture," remarked his partner.

"It's the least I could do," replied the first, heading on toward the next hole, "after 35 years of marriage."

They Never Forget

Wife to inebriated husband: "If it were the first time, Clyde, I could forgive you. But this has happened before in 1954!"

Chapter 14

Man's Better Half?

"I have bursts of being a lady, but it doesn't last long."

—*Shelley Winters*

- A tramp knocked at the door of a quaint old country inn named "George and the Dragon." A woman opened the door.

"Could I trouble you for something to eat?" the tramp pleaded.

"No!" roared the woman as she slammed the door in his face.

A moment later, he knocked again, and the same woman answered the door. "Now what?" she screamed.

"I was just wondering," said the tramp, "if I might have a word with George?"

- Two farmers were commiserating about their wives. "Do you know why it's called PMS?" asked one.

"Got me," the other replied. "Why?"

"Because the name 'Mad Cow Disease' was already taken!"

- A wife was talking to her husband about their 25th wedding anniversary: "This year, let's give each other sensible gifts," she suggested, "like neckties and mink coats."

- "Just think," said the man reading his magazine, "it says here that over 5,000 camels are used each year to make paint brushes!"

"Incredible," answered the woman, impressed. "Isn't it amazing what they can teach animals these days?"

- Farmer Ed was having yet another argument with his wife.

"I'm willing to meet you halfway," his wife offered. "I'll admit I'm wrong if you'll admit I'm right."

- Mrs. Anderson was talking to the manager of the bank about opening an account. "I'd like to open a joint account," she explained, "with someone who has a lot of money."

● An ag banker and his wife were eating at a fancy restaurant when it was held up. The stick-up men went from table to table, looking for valuables to swipe. When they reached the banker's table, one of the crooks noticed the woman's sparkling diamond ring. He walked over, studied it for a second and said, "Ha. This is a phony—a piece of glass."

"A piece of glass!" the wife screamed. "You obviously don't know the first thing about fine jewelry. This ring cost $15,000."

"Fine," the crook said, "have it your way. Hand over the ring."

● A farm wife running errands in town was questioned by her husband after her car rolled into another car at the curb. "Why didn't you set your emergency brake?" he asked her.

"Emergency!" said the farm woman. "Since when is mailing a letter an emergency?"

● A lonely farmer was talking with his friend. "It's terrible growing old alone," he grumbled.

"But you have your wife," the friend comforted.

"My wife hasn't celebrated a birthday in nine years."

● "Good news, dear," called the husband as he entered the house. "I picked up two tickets for the theater on the way home from work."

"Oh, that's wonderful," said the wife, "I'll start dressing right away."

"Good idea," he said. "The tickets are for tomorrow night."

● Lydia pushed back her chair from the table and announced happily, "Guess what? I finally finished that jigsaw puzzle I was working on, and it only took me five weeks!"

Her husband looked over his newspaper and said, "Five weeks? That doesn't sound like much of an accomplishment. Why are you so excited?"

"Well," she explained, still smiling, "it said on the box, 'Three to five years!'"

● "Dear, you'll never know what I went through to buy you your birthday present," the wife said with a sigh.

"Come on, tell me," the husband urged her.

"All right," she replied. "Your pockets."

● The 20 best years of a woman's life are between 35 and 36.

● Two farm wives were shopping at the grocery together.

"I'm really into homestyle meals," one said. "Aren't you?"

"Oh, sure," the other admitted. "If the label on the can doesn't say, 'homestyle,' I don't buy it."

● "Gee, Myrtle, the bank returned my check," Helen said.

"Are you ever lucky!" Myrtle cried. "What are you going to buy with it next?"

● "What is your age?" asked the trial lawyer, "and remember, you are under oath."

"31 and some months," the woman answered proudly.

"How many months?"

Her answer, not quite as proud as the first: "Uh, 108"

● "Recycling is nothing new," Harvey told his co-worker. "In our house, it's called a casserole."

● Farmer Ben was getting agitated. "Are you ready to go yet, dear?" he asked as nicely as he could.

"I wish you would stop nagging me, honey," his wife shot back. "I told you an hour ago, I'll be ready in a few minutes."

● A young bride was describing the wonderful qualities of her husband to her friend.

"John is just the most generous man in the whole world," she beamed. "He gives me everything credit can buy."

● "What would you like for your birthday?" Farmer Todd asked his wife.

"I'd like something that's hard to break," she replied.

"You mean something like Tupperware?" he asked.

"No," she said, "more like a hundred dollar bill."

● There's a farmer down the road who saved thousands of dollars by performing his own plastic surgery.

He cut up all his wife's credit cards.

● Why Man's Best Friend Is Really Better To Have Around Than A Farm Woman

— You never have to wait for a dog; he's ready to go 24 hours a day.
— Dogs rarely outlive you.
— Dogs never want foot massages.
— Dogs find it amusing when you're drunk.

— Dogs have no use for flowers, candy or jewelry.
— Dogs never expect presents.
— Dogs don't hate their bodies.
— Dogs don't let magazine articles run their lives.
— Dogs agree that you have to raise your voice to get your point across.
— Dogs understand that instincts are better than asking for directions.
— Dogs like beer.
— Dogs never try to analyze the relationship.
— If a dog is gorgeous, other dogs don't hate it.
— Dogs like it when you leave lots of things on the floor.
— Dogs don't mind if you give their offspring away.
— Dogs love it when your friends stop by.
— The later you are, the happier a dog is to see you.
— A dog's time in the bathroom is confined to a quick drink.
— Dogs think you sing great.
— Dogs don't care if you use their shampoo.

● "I never repeat gossip," Valerie whispered to her friend. "So, please listen carefully the first time."

CROW BAR

● A working mother and her young son were shopping in a supermarket. The young child, trying to be helpful, picked up a package and dropped it into their cart.

"No, no," protested the mother, pulling the item out of the cart, "go put this back. You have to cook this."

● A newly settled farm woman wrote her best friend back in Arkansas.

"My sister and I aren't exactly lonely here in North Dakota because we have each other to talk to," she wrote.

"The problem is, though, it gets a little boring without another woman to talk about."

● As two liberated woman were talking over coffee one day, their conversation drifted off talk shows and onto domestics. "I had a cookbook once," said one, "but I could never do anything with it."

"Too much fancy work involved in it, huh?" replied her friend.

"You know it," answered the woman. "Every one of the recipes began the same way—'Take a clean dish...'—and that made up my mind."

● Stop praising a woman and she thinks you don't love her anymore; shower her with compliments and she'll think she's too good for you.

● "I can keep a secret just as well as any one," Trudy told her friend. "It's the person they tell it to that can't."

● A woman motorist was driving along a country road with her husband when they noticed a couple of repair men climbing telephone poles.

"What do you know," the man said, "they must have known you were coming."

● The tall, dignified farmer joined the crowd in front of the bargain table, in a vain attempt to buy a very special pair of gloves for his wife's birthday. He inched his way patiently but was buffeted here and there by the women around him and made no progress.

Suddenly, he lowered his head, stretched out his arms and barged through the crowd.

"Can't you act like a gentleman?" asked a cold feminine voice at his elbow.

"Listen, lady," replied the perturbed farmer, still inching forward, "I've been acting like a gentleman for the past 20 minutes. From now on, I'm going to act like a lady!"

Chapter 15

The Man Behind The Woman

"Man: a curious mammal who buys football tickets in June and his wife's Christmas present on December 24."

- Farmer Jed was trying to get himself out of the dog house with his wife.

"And what little present shall I buy for the one I love best?" he cooed.

His wife was quick with her response, "How's about a big stogie and a six pack to go with it?"

- The woman's work that is never done is getting her husband to help around the house.

- The middle-age man was shuffling along, bent over at the waist, as his wife helped him into the doctor's waiting room.

A woman in the office viewed the scene in sympathy. "Arthritis with complications?" she asked.

The wife shook her head. "Do-it-yourself," she explained, "with concrete blocks."

- Men are the greatest hunters. But women find more parking spaces.

- Two motorists met on a bridge too narrow for both cars to pass.

"I never back up for an idiot!" yelled the man in the flashy car.

"That's all right," hollered the woman driver as she shifted into reverse. "I will."

- "The average man," Farmer Hal read aloud from the newspaper, "has 66 pounds of muscle, 40 pounds of bone, 50 pounds of fat and about three pounds of brain."

"Well," replied his wife, "that explains a lot."

- A husband is a man who expects his wife to be perfect and to understand why he isn't.

- Farmer Stu and his wife were returning home after a night in town.
"Did you see how pleased Mrs. Smith was when I told her she didn't look a day older than her daughter?"
"No, dear," his wife said, "I was too busy watching the expression on her daughter's face."

- "Even though times have changed a lot, I still can't help but believe the woman's place is in the home," proclaimed Ted.
"That's right!" agreed Carl. "And I expect to find her there immediately after she gets off work."

- The farm sports fan, who stayed glued to his TV set every weekend, was confronted by his wife.
"You love football more than you do me," she sobbed.
Trying to calm his wife, he replied, "Don't feel so bad, honey, I still love you more than basketball."

- What's a farmer's idea of helping with the housework?
Lifting his legs so his wife can vacuum.

- The farmer visited the local bookstore. "Have you got a book entitled, Man, the Master?" he asked the female clerk.
The salesperson replied, "You'll find that one in the fiction section."

- "That's my husband's ashes in the vase on the mantle," Elizabeth remarked.
"I'm so sorry to hear he's dead," her guest replied.
"Oh, he's not dead," the woman commented. "He's just too lazy to find an ash tray."

- How many men does it take to change a roll of toilet paper?
Nobody knows; it's never happened.

- "My Ned is a do-it-himself kind of guy," Lucy noted to her friend.
"That must come in handy," the friend replied.
"Oh, sure," Lucy admitted, "like when the dishwasher broke, I wanted to call a repairman, but he insisted on calling one himself."

- "A man's life," sighed Joe, "is 20 years of having his mother ask him where he's going, 40 years of having his wife ask the same question, and at the end, the mourners are wondering, too!"

● Jill and Carole were chatting over lunch one afternoon. "Did you hear that the thinnest book ever written was recently released?" Jill asked.

"Interesting," remarked Carole. "What's the title of this book?"

"What Men Know About Women," Jill replied.

● Marie had been on a dinner date with Norm the night before and was recapping the evening's events for her girlfriend.

"I tell ya, Florence, that man had nothing to say," she sighed. "And I had to listen to him the whole night to find that out."

● Considering that about only one man in a million is considered genius enough to understand all the world's problems, it's darn funny how you keep running into him all the time.

● Two farm couples were having dinner together when one of the wives began to sing her husband's praises. "My Harold never said an unkind word about anybody," Rita bragged as her husband reached across the table for the bread basket.

"That's because he never talks about anybody but himself," grumbled the other wife.

● Betty's teenage daughter was studying for her Theology exam. Peeking over her notebook, the daughter asked her mother, "Why exactly did the Israelites wander 40 years in the desert?"

"Because even back then, dear," Betty replied, "the men refused to stop and ask for directions."

● A farmer had been told that he needed a brain transplant and asked his doctor, a female, how much it would set him back. The doctor said she had a nice brain available from a young man and that it would cost $40,000. The farmer told the doctor that was really out of his price range.

Then, the doctor said she had a brain from a woman available for only $20,000. The farmer was curious as to why there was such a price difference between the two brains.

"Well," the doctor explained, "the woman's brain comes used."

● Rex had just returned from a fishing trip with the guys.

"Wait'll you see the big bass I caught, Mary!" exalted the happy angler. "It's a beauty!" Despite his glee, the farmer's face suddenly grew solemn. "But, honey," he continued, "although the fishing trip was fun, I really missed you. I'm so glad to be home. I'm just not happy when I'm away from you, sweetheart."

The wife looked her husband squarely in the face. "I'm not cleaning it."

Who's The Boss?

Chapter 16

"A husband is a man who thinks he bosses the house—when in reality, he only houses the boss."

● The henpecked farmer talked to his doctor and told him of his recurring nightmare. "Every night," he said, "I dream I'm shipwrecked on a tropical paradise with 12 beautiful wives."

"What's so nightmarish about that?" the doctor asked.

"Have you ever tried cooking for 12 women?" the farmer shrieked.

● The father of five children had won a toy at a raffle. He called the kids together and asked, "Who should have this present? Who is the most obedient? Who never talks back to Mommy? Who does everything she says?"

Five small voices answered in unison: "You play with it, Daddy!"

● A young woman was explaining to a friend why she had decided to marry one man rather than another. "When I was with Alfred," she said, "I thought he was the smartest person in the whole world."

"Then why didn't you choose him?" the bewildered friend asked.

"Because when I'm with Arnold, I think I'm the cleverest person in the world."

● Orville's wife was getting tired of the broken kitchen sink. "Orville," she said to her dozing husband on the couch, "the sink needs fixin', and you'll sleep much better after you've fixed it."

"How's that?" he grumbled.

"'Cause, I'm gonna keep waking you up until you do."

● A dish towel will certainly wipe a contented look off a married man's face.

● The farmer was out shopping for a new farm vehicle, but the pushy salesman was trying to talk him into a new car. "Just think, what would your wife say if you took her for a ride in a new car?" the salesman asserted.

The man replied, "Well, she'd say, 'Hey! Look out for that truck!' and 'Watch where you're going!' and things like that."

● A perfect wife is one who helps the husband with the dishes.

● Two farm wives were catching up on things over a cup of coffee. One woman, who was renowned in the community for her expansive remodeling, was asked by her friend if she had done any home improvements over the summer.

"Oh yes, the best one yet," she replied. "I kicked Eugene out of the house."

● Two farm women were talking over afternoon tea about their husbands in the usual manner.

"Henry is perfectly helpless without me," Martha said.

"Donald is the same way," said the other. "I don't know what would become of him if I went away for a week!"

"Isn't that the truth!" sighed the first. "Sometimes I think my husband is a child the way I have to look after him. You know, whenever he is sewing on buttons, mending his pants, or even darning his socks, I always have to thread the needle for him!"

● One day an elderly couple drove to visit their married son and his wife.

When they arrived, the son greeted his father and noticed how haggard he looked.

"Dad, did you do all the driving?" he asked.

"No," Dad replied, "Your mother did. I just sat behind the wheel."

● "I'm sorry to tell you, but your wife's mind is completely gone," the doctor told Clint.

"Well, I'm not surprised," he replied. "She's been giving it to me a piece at a time for 20 years now."

● "Why was Solomon the wisest man in the world?" the preacher asked his congregation.

Judy leaned over to her husband and whispered, "Because he had so many wives to advise him."

● It's easy to spot the husband who won't admit he's hen-pecked. He smokes a big cigar while he washes dishes.

● A farmer who suffered from bursitis was told by his doctor to apply heat for relief.

"But, Doctor," objected the patient, "my wife says it's better to use cold packs."

Replied the doctor, "Well, you tell your wife that my wife thinks heat is better."

● "I wouldn't say my wife always gets her own way," the farmer told his friend, "but she writes her diary entries a week ahead of time."

● "My wife and I have been married for 28 years," Woody told his guests, "and every time I open my mouth, she still corrects me."

"29 years," sang the voice from the kitchen.

● Martha's daughter had come to her mother for guidance on the subject of matrimony.

"You'd better ask your father's advice," she told her daughter. "He made a much smarter marriage decision than I did."

● Russell and Clyde were swapping marriage stories.

"Do you still feel the same toward your wife as you did before you were married?" Russell asked his buddy.

"Just the same," Clyde replied. "I remember when I first fell in love with Mary. I would lean over the fence in front of her house and gaze at her shadow on the living room curtain, afraid to go in.

"And now," he continued, feigning a shiver, "When I come home at night, I feel exactly the same way."

● Old Ezra was celebrating his 100th birthday and everybody commented on how well preserved he appeared.

"I will tell you the secret of success," he started. "My wife and I were married 75 years ago. On our wedding night we made a solemn pledge that whenever we had a fight, the one who was proved wrong would go out and take a walk. Thus, I have been in the open air almost continually for all these years."

● It doesn't matter how often a married man changes his job; he still winds up with the same boss.

Clash Of The Spouses

"Marriage is like a violin. After the beautiful music is over, the strings are still attached."

● Rose confronted her husband over breakfast: "You know you swore at me last night in your sleep?"

To which the husband replied, "Who was sleeping?"

● Farmer Vern waited until his wife had gone to bed, then came in saying, "Here's a glass of water and a couple of aspirins for you."

"But, I don't have a headache," she replied.

"Gotcha!" Vern exclaimed.

● Wayne and Edith were having it out again.

"When we were first married, I was really happy," Wayne said. "I'd come home from a hard day in the fields; the dog would race around barking and you would bring me my slippers.

"But now, everything's changed!" Wayne complained. "When I come home these days, the dog brings me my slippers and you're the one who barks at me!"

"Why are you complaining?" Edith snapped back. "You're still getting the same service, aren't you?"

● Farmer Jones was having another fight with his wife over finances. "You accuse me of spending too much money on the farm?" she asked. When did I ever make a useless purchase?"

"Why, there's that fire extinguisher you bought a year ago," his wife argued. "We've never used it once."

● A couple, who had been married for 25 years, were driving along a country road and came up behind another car in which they saw two people sitting very close together behind the steering wheel.

"Pa," the wife said, "see that couple? Now, why don't we do that anymore?"

"Well, Ma," her husband replied, "I ain't moved."

● "I think you've had enough, buddy," the bartender told one of his regular customers.

"Aw, c'mon," the man slurred. "I just lost my wife."

"Sorry, guy," said the bartender. "I didn't know. That must have been hard."

"Hard?" the man said. "It was nearly impossible!"

● A farmer found a box in his wife's closet that contained four eggs and $1,700 all in $1 bills.

He took the box out of the closet and asked his wife why she's collecting them.

She said, "Well, dear, every time during our marriage that I found our time in bed unfulfilling, I put one egg in the box."

Realizing that they have been married for 35 years, the farmer felt pretty good about himself.

"But what about the $1,700 in $1 bills," the farmer asked his wife.

"Well," she admitted, "every time I get a dozen eggs, I sell them for $1."

● Two ranchers were drowning their sorrows. "I've had bad luck with both my wives," said one rancher to the other.

"How come?" his friend asked.

The rancher replied, "Well, the first one ran off with another man."

"That's too bad," consoled his friend. "What about the second one?"

"She didn't," came the reply.

● I grew up in a farm family that was so big that I never slept alone until after I was married!

● "You know, Herb," the wife said, "Playing chess reminds me of the days when we were dating."

"How so?" asked Herb. "We never played chess back then, Mary."

"No," said the wife, "but even then it took you two hours to make a move."

● Two farmers were discussing buying birthday presents for their wives.

"I just don't know what to get Val," Leroy said. "Got any ideas?"

"Well, last year I got Katherine a cemetery plot," Earl replied.

"What are you going to get her this year?" Leroy asked.

"Nothing," Earl answered. "She didn't use what I gave her last year."

● After listening to her husband's tirade, the wife remarked, "You may not have had a happy childhood, but you're certainly having a long one!"

● "Where is all the grocery money I give you going to?" barked the farmer at his wife.

"Turn sideways and look in the mirror," she snapped.

● A farm couple was arguing in their living room. "You know," began the wife, "I was a fool when I married you."

The husband replied with a grin, "Yes, dear, but I was so smitten, I didn't notice."

● "You know," Farmer Newt groaned, "I think the spark has fizzled from my marriage."

"Aw, c'mon, Newt," Ralph encouraged, "Why do you say that?"

"Well," Newt said, "last night I asked my Helen to wear something see-through to bed, and she did—a hair net."

GOAT TEA

● A farmer and his wife went for a stroll in town and noticed a couple kissing passionately on their porch.

"Why can't you do that?" asked the wife.

"Don't be silly," her husband replied. "I don't even know that woman."

● The man and his wife were having a knock-down, drag-out fight over money.

"You rat!" she screamed, "Before we were married, you told me you were well off!"

"I was!" he hollered, "but I didn't realize how well off!"

● A dejected Martha walked into the beauty parlor and took a seat. "Well," asked her friend, "what did the marriage counselor say?"

Martha replied, "She said we needed more magic in our relationship."

"That's great," said Bertha. "So why are you so down?"

"Because," Martha said, "Stanley decided to disappear."

● Old Clarence was reprimanding his wife. "I think, dear," he said sternly, "that you fib a little on occasion."

"Well, dear," she replied, "I think it's a wife's duty to speak well of her husband from time to time."

● "Herbert's asked me for a divorce!" Helen sobbed to her friend. "We've had nine children together, and now I find out that he never even loved me."

"There, there, Helen," her friend responded. "Just imagine the fix you'd be in today if he did love you."

● Two married farm hands, Larry and Burt, were out last night. The next day, Larry was fine, but Burt was a wreck. Larry asked what happened.

"I cut off the engine and glided into the garage, took off my shoes, tiptoed into the house, undressed in the kitchen and tried to slip into bed silently," he said. "Trudy woke up and nearly killed me!"

"Aw, man, you did it all wrong," Larry explained. "See, I blew the horn, slammed the back door, came in singing, turned on the bedroom light and said, 'Honey! How about a roll in the hay?' Nellie pretended she was asleep!"

● "After I shave in the morning, I always feel 10 years younger," Farmer Ned commented as he walked out of the bathroom.

"Well, dear," his wife said with a grin, "why don't you try shaving before you go to bed at night?"

Cheatin' Hearts

"When a man brings his wife flowers for no reason—there's a reason."

—Molly McGee

● "My wife wants to get romantic in the backseat of our car," Lester confessed to his friend.

"Wow, that's great. But why the long face?" the friend asked.

Lester said, "She wants me to drive."

● A traveling salesman walked up to the hotel counter, escorted by a very beautiful, young woman.

"Have you a room for me and my wife?" the man asked the clerk.

"All I have left is a single," the clerk answered.

"Will that be all right with you, dearest?" the salesman asked the clinging blonde on his arm.

"Whatever you say, mister," she said with a smile.

● A country doctor and his wife were having it out at the breakfast table. "You aren't so great in bed either!" the doctor shouted and stormed off to work.

By midmorning, he decided to make amends and phoned home. After a long time, his wife finally picked up the phone.

"What took you so long?" the husband asked.

"I was in bed," the wife responded.

"What were you doing in bed this late?"

Putting it in terms he could understand, the wife smugly answered, "Getting a second opinion, dear."

● When a newly married man looks happy, we know why. But when a 10-year married man looks happy, we wonder why.

● The farmer tore into his bedroom knowing full well that he would find his wife in bed with another man. "You miserable woman!" the farmer yelled as his face turned redder and redder. "I know everything!"

"Now don't exaggerate," the wife coolly replied. "When was the Battle of Waterloo?"

● Wife: A woman who spends the first part of her life looking for a husband and the last part wondering where he is.

● There was trouble in the Garden of Eden. Adam had been getting home later and later each night, missing supper several nights in a row. Needless to say, Eve was fuming and questioned him as to his whereabouts. Adam said he was out tending the garden, but Eve still was suspicious. "You've got another woman, don't you?" she accused.

"Nonsense, Eve, you know there are only the two of us," Adam told her and went to bed.

During the night, Adam awoke to find Eve tapping on his chest. "What are you doing?" he asked her.

"Just counting your ribs," came Eve's reply.

● "I understand you left your husband because he said you were a lousy lover," Nora said to her neighbor.

"Not exactly," corrected Maria. "I left him because he knew the difference."

● Women are to blame for most of the lying men do. They insist on asking questions!

● Farmer Jones was walking home one evening when a stranger rushed up and hit him in the jaw. "That's for messin' around with my wife, Harris!" the stranger yelled.

Jones picked himself up off the street and stared at the man in amazement. Suddenly, a broad grin spread over his face, and then he roared with laughter.

"What are laughing about, Harris?" exclaimed the attacker. "I just knocked you down."

"The joke's on you, pal," chortled Jones. "I'm not Harris."

● A dramatic actor, whose habit was to embellish his speech with fancy phrases, returned home early one afternoon. The maid met him in the hall and asked if he was looking for his wife.

"Yes," he replied in stentorian tones. "I seek my dearest friend and severest critic."

"Well," she answered wryly, "Your severest critic is in the bedroom and your best friend just jumped out the window."

● Grandpa observes that "too many girls are looking for a husband when they should be looking for a single man."

● The angry wife called on her attorney and announced she wanted to sue her husband for divorce.

"On what grounds?" asked the attorney.

"Bigamy," she replied tersely. "I'll show him he can't have his Kate and Edith, too."

● Two farmers were having a cup of coffee at the town diner. One said to the other, "I have an awful headache. I just don't know what to do to get rid of it."

His friend answered, "Well, do you know what I do when I have a bad headache? I put my head on my wife's lap. She rubs my forehead and sings to me and my headache just goes away. It really works—you should try it."

The first farmer jumped up and said, "Great! Do you think she's home right now?"

● A farmer named Jones insisted on having one night out a week, alone. Every Tuesday night he left the house for some fun. One particular Tuesday night the farmer went out and didn't come back for seven years.

When he did finally return, his wife was so happy that she phoned all her friends.

"What are you doing?" Jones demanded.

"I'm arranging a welcome home party for you tonight," the wife answered sweetly.

"What? Are you crazy?" yelled Jones. "On a Tuesday?"

● The church service was proceeding successfully when an attractive young widow, seated in the balcony, leaned too far over the railing and fell. Her dress caught in the chandelier, and she was left hanging helplessly in mid-air.

The minister noticed her undignified position and thundered, "Any person who turns to look will be stricken blind."

A man whispered to his neighbor next to him and whispered, "I reckon I'm going to risk at least the left eye."

● "I don't like to pry into your affairs," admitted the farm wife to her husband, "but something's been bothering me for days."

"So," said the husband, "tell me all about it."

"Well, you got a letter last Friday," the wife began, "and it was perfumed. It was in a girl's handwriting, and I saw you open it and then break out in a sweat. You turned white. Your hands trembled. For goodness sake, who was it from and what did it say?"

"Oh, that," replied the husband very calmly. "I decided it was best for both of us not to talk about it at the time. I've been trying to think of the best way of discussing it with you without causing an explosion."

"My God!" screamed the woman. "Tell me who it was from and what it said!"

"OK," said the husband. "It was from your dress shop. It said you owe them $360."

- Farm woman on the telephone: "Hello, is this Fidelity Insurance? I want to have my husband's fidelity insured."

- A farmer, a co-op manager and a tractor salesman were discussing whether it was better to have a wife or a mistress.

"Guys, nothing beats being married," the farmer said. "I enjoy building a solid foundation for an enduring relationship."

"A mistress," the co-op manger countered, "is my source of mystery and passion."

"I prefer both," said the salesman to his startled friends. "Each assumes you are spending time with the other, and you can go the office and get some work done."

- Newspaper reporters sometimes risk their lives to bring home a story. Other men risk their lives with the stories they bring home.

- "I'm sorry, Mrs. Hackney," the doctor said, "but I can't cure your husband of talking in his sleep."

"But, Doctor," the farm wife said, "I don't want him cured. I just want you to give him something that will make him talk more clearly."

- "A fine time to come in," stormed Betsy as her husband stumbled in at 4 a.m. "I want an explanation and I want the truth!"

"Make up your mind, dear," the farmer replied. "You can't have both."

- "For months," said the free-spirited woman, "I wondered where my husband spent his evenings."

"And then what happened?" asked her friend.

"Well, one evening I went home. And what'll you know—there he was!"

- Farmer Jerrod's attempt at sneaking in his house at 3 a.m. was unsuccessful.

"You'll never guess where I've been tonight," he said, checking his collar for lipstick.

"Oh, yes I can," his wife replied through clenched teeth. "But go ahead with your story anyway."

● Mrs. Allen had a bone to pick with her local post office.

"I wish to complain about the service," she told the man at the window.

"What's the trouble, lady?" the clerk asked.

"I don't know what kind of business you think you're running, but it's all screwed up," she nagged, whipping out an envelope from her purse. "My husband is in Chicago on business and this letter he sent me is post-marked Las Vegas!"

● "I just heard your husband is in the hospital," gushed the wife's friend. "What happened?"

"It's his knee," answered the wife. "I found a redhead on it."

● "Some people are funny," a local philosopher observed.

"You just found that out?" his friend asked. "Why the sudden revelation?"

"Well, I know a guy who hasn't kissed his wife for 10 years," the philosopher explained. "Then he goes out and shoots a guy who did!"

● Have you heard about the new glasses developed for men to watch miniskirts?

They're known as thigh-focals.

● In the middle of the night, Farmer Gus' wife began to dream she was in the arms of another man. Then she dreamed she saw her husband coming after her, and still asleep, yelled out, "Oh, no! My husband!"

The slumbering Gus, awakened by the cry, jumped out the window.

● "Don't you worry about you husband running after other women?" asked Gertrude.

"No—he's like a dog chasing cars," her friend replied. "He wouldn't know what to do if he caught one."

● A nurse in a mental hospital was relaying a message to the doctor. "There's a lady outside who wants to know if we've lost any female inmates," she said.

The doctor replied, "That's odd. Why?"

The nurse said, "She says some woman ran off with her husband."

● "I've been married for 25 years and have never stopped being romantic," the country agent told his co-worker.

"Yeah, but if your wife ever found out," razzed his friend, "she'd break you neck."

● Two farm women were talking about long hair on men.

"Personally," said one, "I think long hair makes a man look intelligent."

"Oh, I don't know about that," said the other woman. "I picked a long hair off my husband's coat, and he looked mighty foolish to me."

● "Isn't that nice?" remarked the crop consultant's female colleague. "He phones his wife several times a day."

"Yeah," replied another, "but I know several men who do the same thing."

● The wife was terribly jealous. Evening after evening she subjected her husband to a searching inspection. When she would find even a single hair on his coat, there would be a terrible scene.

One night, she found nothing. "So!" she screamed. "Now it's a bald-headed woman!"

● The rural schoolteacher wrote this note to Ryan's mother: "Ryan is a good student, but I must find a way to take his mind off the girls."

Ryan's mother sent back this reply: "If you find a way, let me know. I'm having the same trouble with his father."

● A traveling feed dealer stopped for room and board at a farmhouse. Before he could get a second helping of the woman's delicious mashed potatoes, the farmer whipped them off the table and put them in the refrigerator.

After supper, the farmer went out to shut the chicken coop. The farmer's wife slid over beside the salesman, batted her eyelashes, puckered up and said in a breathy voice, "Now's your chance."

So, the salesman got up and ate the rest of the mashed potatoes.

● Two farm wives were sharing an afternoon tea.

"My husband and I are always having arguments," the first wife complained. "I don't think we have a single thing in common."

"Same here," said the second. "No, I'm wrong. I like a night out with the girls and so does he."

● Two farmers were drinking at the bar. "Why the sad look?" Milo asked.

"Aw, it's nothin'. I can't tell you about it," replied Otis.

"Here, have a another drink, then maybe you'll feel like talkin'," Milo said.

Sure enough, after downing a couple, old Otis loosened up.

"There you go," encouraged Milo. "Now, tell me, what's troublin' ya?"

Otis put his arm around his friend and said, "Well, buddy-boy. I hate to tell you this, but it's your wife."

"My wife?" asked Milo.

Ottis responded, "Yup. I'm afraid she's cheatin' on us."

● You know your wife is sneaking around when you notice your dog carrying your pipe and slippers to a house down the block.

● Testimony on an Alaskan bigamy trial revealed that the defendant had one wife in Nome, a second in Juneau and a third in Fairbanks.

"How could you do such a thing?" the judge asked.

"Fast dog team, " the defendant boasted.

● "Last night I heard my wife talking dirty," admitted Archie.

"Wow, after all these years. You're a lucky guy, Arch," his pal remarked.

"Real lucky," Archie grumbled. "When I looked to see where she was, I caught her on the phone."

● "St. Peter, is my husband here?" the woman asked. "His name is Smith."

"We have a lot of Smiths here," the saint replied. "Could you be a little more specific?"

"Joe Smith," said the woman.

"We have a lot of Joe Smiths," said St. Peter. "Is there any other identification?"

"Well, when he died, he said that if I was ever unfaithful to him, he would turn over in his grave," explained the wife.

"Oh, you mean Joe "Pinwheel" Smith! Yes, he's here," said St. Peter.

Where Have You Been?

Then there was the woman who applied for a divorce. She said her husband was very careless about his appearance—he hadn't shown up at home in nearly four days.

Chapter 19

You Can Pick Your Friends, But You Can't Pick Your Family

"I wanted to do something nice so I bought my mother-in-law a chair. Now they won't let me plug it in."

—Henny Youngman

- The airlines weren't given enough credit for the job they did during the holidays. Without their delays, we'd have to put up with our visiting relatives from the city a lot longer.

- "You've got a pretty place here, Nate," the departing mother-in-law stated, "but it looks a bit bare."

  "Oh, it's because the trees are rather young," Nate said, showing his guest to her car. "I hope they'll have grown to a good size before you come again."

- What's the penalty for bigamy?

  Two mothers-in-law.

- "I baked two kinds of cookies today," the mother-in-law told her daughter's husband, tired and hungry from a hard day's work on the farm. "Would you like to take your pick?"

  "No," the son-in-law replied, "I think my hammer will work just fine."

- A rich banker died and his family met for the reading of the will. The banker left $180,000 to his wife, $75,000 to each of his brothers and $50,000 to each of his sisters.

  The will went on, "And to my nephew, Ralph, who always hinted that he wanted to be mentioned in my will... 'Hello, Ralph.'"

- Two farmers were commiserating about their mother-in-laws. "I wouldn't say my mother-in-law is bad, it's just that she's so nearsighted," admitted one. “The other day she nagged a coat hanger for almost an hour.”

● Married 10 years, the farm couple decided to live it up on a second honeymoon. While talking over their plans one night, the husband kept glancing into the living room where a little old lady sat, knitting.

“The only thing that worries me,” he said, “is what are we going to do about your mother when we go away?”

“My mother!” his startled wife said. “I thought she was your mother.”

● Farmer Bert’s in-laws were visiting from out of town. To escape the unfortunate situation, if even for only an afternoon, Bert decided to make that doctor’s appointment he had been putting off for so long.

After a long, thorough exam, Bert’s doctor, a long time friend, gave him some smart advice.

“Bert,” the doc advised, “You’d better stop drinking so much. It’s affecting your hearing.”

“I’d rather not, Doc,” the farmer replied. “The stuff I’ve been drinking is a lot better than the stuff I’ve been hearing lately.”

● The problem with the gene pool is that there is no lifeguard.

● “Tell me, Farmer Ted, is there any insanity in your family?” the doctor asked.

“There must be,” answered Ted, “they keep writing me for money.”

● Don’t try to make your visiting relatives feel at home. If they had wanted to feel at home, they would have stayed there.

● Farmer Carl was complaining about his mouthy mother-in-law.

“The woman will not stop nagging me!” he complained. “I love my wife, but she’s quite a bit of baggage!”

“You know, Carl,” his friend said, “I think mothers-in-law are a lot like seeds. You don’t want them, but they come with the tomatoes.”

● My brother-in-law is not very bright. He once sat up all night, wondering where the sun comes from. Finally, it dawned on him.

● How are grandparents and harps alike?

Both are unforgiving and hard to get into your car.

● Two brothers grew up on a Minnesota farm. Reaching adulthood, one decided to stay on the farm and work the land with his father. The other son left for Minneapolis to become a banker.

One day, the banker received a call from his brother back on the farm. "Dad's died," he said.

"Oh, I'm swapped with work," the banker replied. "There's no way I can get back for the funeral. But, I'll pay for all the funeral expenses."

The farmer agreed. "But, make sure Dad goes out in style, OK?" he added.

Sure enough, the checks started rolling in to cover the burial fees. The farmer received a nice check for $7,800 the first week which made him very happy. By the second week, however, the amounts dwindled down to $130.

The farmer, deeply disturbed, called his brother in the city. "Why the sudden change in money?" the farmer asked.

"Well," the banker explained, "You said you wanted Dad to go out with some class. So I buried him in a rented box."

● Did you hear about the brother-in-law who put on a clean pair of socks every day?

By the end of the week, he couldn't get his shoes on.

● Misers may not be much to live with, but they often make excellent ancestors.

● Ernie's mother-in-law was complaining to her daughter about the ill manners of a friend who had just left. "If that woman yawned once while I was talking, she yawned 10 times!"

"I don't think she was yawning," Ernie chimed in, "I think she wanted to say something."

● The Johnson family was about to sit down to Sunday dinner of ham with all the fixings, when Mrs. Johnson looked out and saw her sister's family turning in the front gate.

"Oh, my goodness," she exclaimed, "here comes company and I'll bet they haven't eaten yet."

"Quick," urged Farmer Johnson, "everybody out on the front porch with toothpicks!"

● Cousin Charlie is tighter than spandex on an elephant: He got married in his own backyard so his chickens could have the rice.

● Trudy's in-laws were gathering their things to leave her house.

"Good night," they said, making their way to the door, "We hope we haven't kept you up too late."

"Not at all," yawned Trudy. "We would have been getting up about now anyway."

● They had been dating for nearly a year already, but he had not proposed yet.

"Georgette," he said, as they were taking a moonlight stroll one evening, "I am-er-going to ask you an important question."

"Oh, James!" she exclaimed. "This is so sudden. Why I-"

"What I want to ask you is this," he interrupted. "What date have you and your mother decided upon for our wedding?"

● Jim and Dan were discussing Jim's night out with his in-laws. "I don't know what my father-in-law does with his money, but I can tell you this," Jim complained, "he sure as hell doesn't carry it with him when we go out for a drink."

"So, did you end up paying for the whole meal?" asked Dan.

"Let's put it this way. Kate's father is so cheap, he would have asked for separate checks at the Last Supper."

Copy Cat Concerns

Little Harry had bought Grandma a book for her birthday and wanted to write a suitable inscription on the inside cover. He racked his brain and suddenly remembered that his father had a book with an inscription of which he was proud. Harry decided to use that one in his grandmother's book.

Grandma was surprised when she opened her book, a Bible, and found neatly written the following phrase: "To Grandma, with compliments of the author."

Chapter 20

The Joys Of Parenthood

"The way I see it, if the kids are still alive when my husband comes home from work, then I've done my job."

—Roseanne

● After experiencing the birth of her sixth child, the young farm woman went to her doctor and asked what she could do to avoid becoming pregnant again.

The doctor rubbed his chin and said, "I would suggest you drink orange juice."

"Before or after?" the young woman asked.

To which the doctor replied, "My God, woman! Instead!"

● The son of a country store manger came home for spring break wearing torn, baggy pants that he was constantly tripping on. "You look like a damn fool!" his father growled at the son.

Just then a friend of the family emerged from the kitchen. Welcoming the son back home for vacation, the neighbor smiled and told him, "Andy, you know, you look more and more like your father every day!"

"You don't say?" remarked Andy. "Dad just got done telling me the same exact thing!"

● "How are Mary's violin lessons going?" asked Farmer Ted.

"Let me just say this," Farmer Hank said, "whoever said practice makes perfect didn't have a child taking violin lessons."

● "So, do you really like being a new mom?" Helen asked her girlfriend, Maggie.

"It's good," Maggie replied through bloodshot eyes. "But people who say they sleep like a baby surely don't have one."

● "Dad, why can't we have wall-to-wall carpeting like the other people?" complained the farmer's son.

"Son, when I was your age," the farmer replied, "we were lucky to have wall-to-wall floors!"

● "Don't bite the hand that feeds you, Jim!" the farmer yelled during an argument with his son.

"Yeah," Jim replied, "it's apt to be dirty!"

● "How did Danny's trip to Doc Smith go?" Mrs. Henshaw asked Mrs. Potter.

"Oh, it went fine. Doc put a small tube in Andy's mouth and told him not to open it for 15 minutes," Mrs. Potter responded.

"Doesn't sound too bad," said Mrs. Henshaw.

"Are you kidding?" asked Mrs. Potter. "It was wonderful. I offered Doc Smith $20 for that little tube!"

● At the monthly 4-H meeting, two mothers were swapping stories. "My kids were giving me such a headache! But I was fine once I followed the bottle of aspirin's instructions," recalled Meg.

The other mom said, "You mean you swallowed two aspirins and it went away?"

"No," explained Meg. "I followed the directions on the bottle. It said, 'Keep away from children.'"

● Junior was constantly getting into trouble, so his dad was surprised when Mom suggested they buy him a new bicycle. "Do you really think it will improve his behavior?" the father asked.

"No," said Mom grimly, "but it'll spread it out over a wider area."

● "I don't enjoy listening to long speeches anymore than you do," bellowed Farmer Ray to his nine-year-old son. "But, unfortunately for you, I do enjoy givin' 'em."

● A young mother of four small children way up in Canada received from her stateside friend a gift of a playpen.

"Thank you so much for the gift," she wrote. "It is wonderful. I sit in it every afternoon and read. The children can't get even close to me."

● "Daddy!" the farm boy exclaimed, racing proudly to his frugal father. "Today I saved a whole dollar!"

"Congratulations, son!" the farmer beamed. "Tell me, how did you save that dollar?"

"Well, I was going to take the bus into town today, but I decided it was too expensive. So, I ran all the way, and saved one dollar!"

"That's a good boy," the farmer encouraged him. "But next time do what I do and run behind a taxi. That way you can save $30!"

● Little Danny was moving his food around on his plate and grimacing.

"Danny!" his mother scolded, "20 years from now you'll be telling some girl what a great cook I was. Now, eat your dinner!"

● A farm couple wanted to have a baby but couldn't. A priest from the local rural parish who called the house said he was going to Rome for a long pilgrimage and that he would light a candle there for their intention to have a baby.

The priest didn't return for another six years. When he did, he called on his friends and asked, "So, do you have any children?"

The woman proudly said, "Why, yes, we have six and another on the way!"

"But where is your husband?" asked the priest.

She replied, "Oh, he's gone to Rome—to put out the candle."

● Little Johnny was in one of his very bad moods. In answer to his mother's demands that he behave himself, he said, "Give me a nickel and then I'll be good!"

"Give you a nickel? Ha!" the mother laughed. "Why, Johnny, you should be good for nothing—like your father."

● The most difficult careers are entrusted primarily to amateurs: citizenship and parenthood.

● During the last days of the Christmas rush at a large department store, a frenzied clerk, overwhelmed by pushing women shoppers, was making out what she hoped would be the last transactions of the day.

As the customer, standing with her four squirmy, young children, gave her name and address, the clerk, pushing her hair up from her damp forehead, remarked, "It's a madhouse, isn't it?"

"No," the customer replied pleasantly. "It may seem like it, but it's a private home."

● A young boy came running up to his dad on the beach, dragging the top half of a bikini along the sand. "Now, be a good boy, Dennis," instructed the father, "and show Daddy exactly where you found it."

● A social worker visited a hillbilly family, the newest addition to her case load.

The woman was a bit shocked after getting the run down on the family history. "Don't any of your nine children have the same father?" she asked.

"Yes'm," came the reply. "I reckon the twins do."

● A 63-year-old farm woman set a record when she gave birth to a son. Shortly after she left the hospital, some of the local woman stopped by her farm to visit the new baby.

But when they asked to see him, the farm wife said, "Let's just visit until he cries."

"Why?" her friends asked.

"Because," she explained, "I forgot where I put him."

● "How is it that your children are so well-behaved compared with those savages of mine?" a farm woman asked her friend. "Why, my boys are always running around with their shirts hanging out. Your boys always keep their shirts inside their pants. What do you do to keep them so tidy?"

"Oh, it's simple," remarked the clever mom. "On the end of each shirt I sewed on a nice piece of lace edging."

● "I just got back from a pleasure trip," Marcy told her neighbor.

"You don't say," the neighbor replied. "Where'd you go?"

"I drove the kids to school."

● A farmer, buying a doll for his little girl's Christmas present, was told by the saleslady, "Here's a lovely doll. You lay her down and she closes her eyes, just like a real little girl."

The sadly experienced father replied, "Lady, it's obvious you've never had a real little girl."

● "How are your children doing in school?" Paula asked her friend over morning coffee.

"Better," replied the weary mother, "but I still go to PTA meetings under an assumed name."

● There should be passage of a new child labor law—one which will prevent the kids from working the old man to death.

● Farmer Tom was reading his son's report card.

"Son, it's too bad they don't give a grade for courage," he said. "You would get an A for bringing this report card home."

● "Remember, dear," the young father told his wife, "you must cultivate patience because the hand that rocks the cradle rules the world."

"Then you come right in here, my dear, and rule the world for a while," she said. "I'm tired."

● "Dad," a young farm boy asked, "what has really great curves, gives you a funny feeling in your stomach and makes you want to jump out of your chair and whistle?"

"Easy, son," replied the father. "Roger Clemens' pitching."

"Thanks, Dad," said the boy as he happily walked away. "I knew there was a reason Mom said I better ask you."

● A farmer and his small son were sitting on a bus heading into town. The boy, his nose pressed against the window, kept asking questions. "What's on that truck, Dad?"

"I dunno," mumbled the farmer, immersed in his paper.

"Where does the rain come from?" said the boy.

"Don't bother me," came the farmer's reply.

"What does that sign say?" asked the son.

Dad said, "Quit pesterin' me!"

A passenger behind them tapped the farmer on the shoulder. "Curious little guy you've got there," she commented.

"Sure," said the farmer. "How else will he learn?"

● The farm couple was making their way through their front door after a dinner and a show.

"Sorry we're coming home so late," they told the babysitter.

"Don't apologize," said the girl. "If I had a boy like yours, I wouldn't be in a hurry to get home either."

● The best time of life for parents is when the kids are too old to cry and too young to drive the car.

● A little old lady, riding on the bus, leaned over and told a woman, "I'd give 10 years of my life to have such fine and nice-looking youngsters as you have."

"That," replied the seasoned mother, "is just what it cost me."

● The day Farmer Phil and his wife brought their first-born home from the hospital, their country doctor, realizing how jittery they were, said reassuringly: "You'll do fine if you just remember two things—keep one end full and the other end dry."

● A proud father phoned the newspaper to report the birth of his triplets. The girl at the news desk didn't quite catch the message. "Will you repeat that?" she asked.

"Not if I can help it," he replied.

● "Has your baby learned to talk yet?" a friend asked the new mother.
"Oh, yes," the women replied. "We're on to the next step already—trying to teach him to be quiet."

● Farmer Chuck was feeling blue and his wife was trying to cheer him up. "Chuck," she said, "look at all that is around you! Missy's doll house is only half finished and Heidi wants you to help her with her driving lessons. By the way, Jeremy's tuba recital is tomorrow and we've got three more kids to put through college.
"And you say you have nothing to live for!"

● The farmer and his small son squeezed themselves into a crowded elevator. A rather large lady turned to the father and said, "Aren't you afraid he'll be squashed?"
"Not at all," replied the farmer, "he bites."

● "Alex! Stop making that nasty face!" his father scolded. "When I was a kid, they told me if I made nasty faces, my face would stay that way."
Alex thought about that one for a second. "So, why didn't you stop?"

● The entire family was assembled for Christmas dinner. All four children were there, each with their newlywed spouses.
Before saying grace, Father made an announcement, "I will give $20,000 to the son or daughter who presents me my first grandchild." With that he bowed his head to pray.
When he looked up again, only he and Mother were left at the table.

● Maxine's 10-year-old son came bounding in from school one day to find his mother in bed, sick.
"Don't you feel well, Mom?" he asked with concern.
"No, honey, I don't," Maxine managed to squeak out.
"Well, don't worry about dinner," he assured his mother. "I can call some of the guys over to help me. With the three of us, I'm sure we could carry you down to the kitchen."

● "Father's Day and Mother's Day are basically the same thing," concluded Jimmy.
"Yeah, sort of," replied Greg. "Except on Father's Day, you buy a much cheaper gift."

● Home is the place where Dad is free to say anything he pleases. No one will pay the slightest bit of attention to him anyway.

Chapter 21

Little Darlin's

"You are only young once, after that you must think of excuses."

● A magazine salesman called the Zimmerman farm and found himself talking to Robbie, one of the Zimmerman children.

"Is your father home?" the salesman inquired.

"No, he's not," answered Robbie.

"Well, how's about your mother?" asked the man.

"She's not around either," Robbie responded.

"Is anyone home?"

"My sister."

"May I talk with her?" the man asked.

Robbie said, "Sure," and went to get her. After a long time, the boy came back to the phone, out of breath.

"I'm sorry, sir," Robbie apologized. "You can't talk to my sister. I can't lift her out of the playpen."

● The school board meeting was interrupted by an announcement. "Are Mr. and Mrs. Miller in the audience?"

Mr. Miller raised his hand to be seen. "Yes, here we are."

"Your little boy is on the phone—he wants to know where your fire extinguisher is."

● Childhood is that wonderful time when all you need to do to lose weight is bathe.

● Two farm boys were talking over their lunch boxes. "Daddy fell into the well in the backyard last week," Freddie began.

"Oh, wow," came his friend's reply. "Is he all right?"

"He must be," said the Freddie. "He stopped yelling for help yesterday."

● "Where were you yesterday, Billy?" asked his friend. "We missed you at the baseball game."

"I fell off a 65 foot ladder while I was helping my dad paint the house," Billy proudly replied.

"Wow! It's amazing you're still alive!"

"Well...," Billy slowly admitted, "I only fell off the first rung."

● "Randy, why were you running up the street?" his mother asked.
"I was trying to stop a fight," Randy answered, trying to catch his breath.
"Oh, my. What a brave young man you are. Who was fighting?"
"Me and another guy."

● "And what did my little angel do all day?" the mother asked, returning from an afternoon in town.
"I played postman, Mommy," the small boy replied, smiling. "I put a letter in every mailbox in the neighborhood. Real letters, too. I found them in your drawer, tied up in pink ribbon."

● A small boy purchased a ticket for the afternoon show at the theater. The manager followed him in and asked, "Son, aren't you supposed to be in school today?"
"I don't have to go to school today," the boy replied proudly. "I have the measles."

● "So, what's new around your house?" the neighbor asked little Betty.
"I don't know," Betty replied sadly. "They spell everything!"

● Stevie came home from school clutching an unexplained $5 bill.
"Where did you get that from?" his mother asked, suspicious.
"Wayne gave it to me for doing him a favor."
A smile crossed Stevie's mother's face. "Oh," she said. "What sort of favor, dear?"
"Well, I was hitting him on the head with a two-by-four and he asked me to stop."

● Farmer Garth was sitting in the armchair one evening, when his little son came in and showed him a new pocket watch.
"I found it in the street," said the son.
"Now, are you sure it was lost?" inquired Garth.
"Of course, it was lost!' replied the youngster. "I saw the man looking for it!"

● "Are your mother and father in?" asked the old lady when the small boy answered the door.
"They was in," said the boy. "But they is out."
"They was in! They was out!" exclaimed the woman. "Where's your grammar?"
"Oh, she's in the kitchen," came the reply.

● Little Mikey was in trouble again. He spent his whole evening sulking in his room until his brother paid him a visit.

"You know," complained Mikey, "Dad must have been a pretty mischievous boy."

"Why do you say that?" asked his brother.

"Because he knows exactly what questions to ask when he wants to know what I've been doing."

● A small boy in a department store was standing near the escalator, watching the moving handrail. "Something wrong, Son?" inquired a clerk.

"Nope," replied the boy. "Just waiting for my gum to come back."

● Little Joey, crying heavily, came in from the barn, where his dad was building a new loft. "Why, Joey, what's the matter?" his mother asked, very concerned.

"D-d-addy hit his f-f-inger with the hammer," sobbed Joey.

"Well, you shouldn't cry about anything like that," comforted his mother. "I would think you might find something like that kind of funny. Why didn't you laugh?"

"I did," cried Joey.

● A farmer stopped to watch a Little League baseball game. "What's the score?" he asked one of the youngsters.

"We're behind 18 to 0."

"Goodness!" exclaimed the man. As he looked over the team he added, "I must say, none of you seem to be very discouraged by the score."

"Why should we?" the boy shrugged. "We haven't even been up to bat yet."

● Farmer Tom came in from the field and found his young son sitting in front of the TV set, clutching a suitcase.

"What's Paul up to?" he asked his wife.

"He's running away from home," she explained. "Just as soon as the show is over."

● A little boy told the salesman he was looking for a birthday gift for his mother and asked to see some cookie jars. At a counter with a large selection of jars, he carefully replaced each lid. His face fell as he came to the last one. He asked, "Aren't there any jars with lids that don't make a noise when you turn 'em?"

● Albert was tapping on his mother's shoulder during the train ride. "Mom, what was the name of the last station we stopped at?" he asked.
"Don't bother me, Freddie," she scolded. "I'm reading."
"I thought you might want to know, Mom," Albert persisted, "because that's where little brother got off."

● "Were you a good, polite girl at church today?" Susie's mother asked.
"Oh, yes," replied the little girl. "A man offered me a big plate of money and I said, 'No, thank you.'"

● "So, Julie, what are you going to give your brother for his birthday?" asked the visiting aunt.
"I don't know," replied Julie, shrugging her shoulders. "I don't think I can top last year's gift."
"Oh, how nice," the aunt commented. "What was it that you gave him last year?"
"Measles."

● Two youngsters were talking about their responsibilities around the farm. "Did Dad promise you something for doing the chores?" asked Mary.
"No," replied Joe, "only if I didn't."

● A farmer took his son to his first football game and later asked him how he enjoyed it. "It seems like an awful lot of trouble over 25 cents," the boy answered.
"What do you mean?" asked his father.
The boy explained, "Everyone in the stands kept yelling, 'Get the quarterback!'"

● "I got tired of hearing that mean old Georgie Franklin saying, 'Sticks and stones may break my bones, but names will never hurt me,'" complained Scott.
"So what did you do, son?" Scott's father asked.
"I hit him with a dictionary!"

● "Eddie," scolded the young boy's mother, "your face is clean, but how did you manage to get your hands so dirty?"
"Washing my face," Eddie replied.

● Officer Sam was out patrolling the area when he spotted a small boy lying unconscious on the sidewalk. Sam approached the boy, and being very concerned, asked the lad what was wrong.

"I was sleeping," the boy said.

"Why on the sidewalk?" Sam inquired.

The boy looked at the officer and simply stated, "Well, that's where I fell asleep."

● A youngster walked into a bank to open an account with $50. The teller smiled and asked him how he had accumulated so much money.

"Selling magazine subscriptions," said the boy.

"You've done very well," said the teller. "Lots of people must have bought them."

"Nope," answered the boy proudly. "Only one family—their dog bit me."

● Kenny looked longingly at his friend's beagle. "My mom won't let me have a dog," he said sadly.

"Maybe you didn't use the right strategy," the friend replied.

"What strategy?" Kenny inquired anxiously.

The friend leaned over and whispered, "Don't ask for the dog. Ask for a baby brother. Then you'll get the dog."

● A little boy caught in mischief was asked by his mother, "Now, how do you expect to get into Heaven, young man?"

The boy thought for a moment and then said, "Well, I figure I'll just run in and out and in and out and keep slamming the door until they say, 'For goodness sake, either come in or stay out.' Then I'll go in."

The School Yard Standoff

A fifth grade boy was finally standing up to the school bully and was now face to face with his nemesis in the school yard. The two proceeded to heckle each other and prance around as a crowd gathered to watch.

After a few minutes, the boy glided over to his friend in the crowd and whispered, "So, I look pretty good, eh? Think I can beat him?"

"Sure," encouraged the friend. "If you keep waving your hands through the air like that he's bound to wind up with pneumonia."

Chapter 22

Kids Call It Like It Is

"Out of the mouths of babes come words Dad never should have said."

● "Travis!" his mother exclaimed after the little boy said a most unsuitable word. "Where on earth did you learn that?"

"From Daddy," Travis replied meekly.

"Well, don't ever let me hear you say it again—it's a bad word," said Mom.

"But why?" Travis asked.

"Never mind—you're too young to know what it means," his mother replied firmly.

"But Mom, I already know what it means," the boy beamed back defiantly. "It means the car has a flat tire!"

● A dairy farmer and his wife were becoming anxious about their four-year-old son, who still had yet to utter his first words. The troubled parents took him to specialists, who said there was nothing wrong with him. But he still didn't talk.

Finally, one morning at breakfast, the four-year-old blurted out, "Mom, the oatmeal is soggy!"

The mother was ecstatic. "You talked! You talked! I'm so thrilled! Why has it taken you so long?"

The boy responded, "Well, up to now I've had nothing to complain about."

● Two farm boys were playing marbles on the playground when a pretty little girl walked by.

"I'll tell you," said one, "when I stop hating girls, that's the one I'm going to stop hating first!"

● A small girl was taken to church for the first time. When everyone knelt down, she whispered, "What are they going to do?"

"They are going to say their prayers," the mother whispered back.

The child looked up in amazement. Then, in a loud voice, she exclaimed, "What? With all their clothes on?"

● A farmer was showing his son, Georgie, the family album and came across a picture of himself and his wife on their wedding day.

"Was that the day Mom came to work for us?" asked Georgie.

● The marble tournament was in full swing at the county fair. One little farm boy missed an easy shot and slipped out a cuss word.

"Franklin!" called the preacher from the spectators' bench. "What do little boys who swear during a game of marbles turn into?"

"Golfers," came the reply.

● "Jimmy," scolded his mother, "stop reaching across the table like that. Don't you have a tongue?"

"Yes," said Jimmy, "but my arm's longer."

● The Davis family had just had triplets. On the way home from the hospital with his parents and new siblings, Little Jimmy was deep in thought.

"What are you thinkin' about, honey?" Jimmy's mother asked.

"I don't know, Mom," Jimmy replied, shaking his head, "We'd better start making some phone calls. They're going to be a lot harder to get rid of than those kittens."

● A young daughter watched her father finish a big Christmas dinner and then loosen his belt.

"Look, Mommy!" she exclaimed. "Dad's just moved his decimal point over two places."

● The Fosters and their daughter, Nancy, were taking a drive into town. Mrs. Foster asked to be dropped off at the beauty parlor.

Watching her mother disappear into the salon, Little Nancy asked, "Daddy, before you got married to Mommy, who told you how to drive?"

● The family cat had just presented the household with a litter of kittens. Junior went down into the cellar to inspect them, and shortly after, his mother heard a medley of howls.

"Don't hurt the kittens, dear," she called out.

"I'm not, Mom," the youngster yelled back up. "I'm carrying 'em real careful—by their stems."

● The 10-year-old daughter of a local veterinarian went into a drug store. She found no one working the soda fountain, so she wandered back to where the druggist was filling prescriptions.

"Can I do something for you?" the man asked.

After quite a pause, the girl asked, "Do you make pregnancy tests for cows?"

"Yes."

"Do you do urine analysis?"

"Why, yes."

"Well, then," the girl said, "would you mind washing your hands and getting me an ice cream cone?"

● A mother was trying to calm her son's fears about surgery. "Ricky, you don't have to be afraid of going to the hospital to have your tonsils out," she said.

"I'm not afraid, Mom," said Ricky. "But I'm not gonna let 'em palm a baby off on me like they did on you. I want a beagle pup."

● "Where did you get that pretty red hair?" a small girl was asked.

"I think I got it from Daddy," she replied. "Mommy still has hers."

● "Hey, Mom," the little boy yelled as he ran into the house. "Dad took me to the zoo again this afternoon, and one of the animals came in and paid him $32.80!"

UDDER CHAOS

● A mother, expecting her third child, told her two small sons that she had a secret to tell them. "Boys," she whispered, "I will soon bring you a little brother or sister."

The little five-year-old joyfully said, "Oh, good, I'm going to go and tell Daddy!"

● Wendy's aunt from the country was visiting at her house. She was a pleasant but plump woman, weighing about 200 pounds. On the night of her arrival, Wendy wanted to stay up later than usual.

"Why, Wendy," her aunt said, "I'm so much older than you, and I go to bed with the chickens every night."

Wendy studied her in silence for a while and then said, "I don't see how you could ever get up on the roost!"

● Little Molly, a six-year-old, complained to her mother, "I've got a stomach ache."

"That's because your stomach is empty," her mother explained. "You'd feel better if you had something in it."

That afternoon the minister stopped by, and in the course of the conversation, remarked that he had been suffering all day from a very bad headache.

Little Molly was alert. "That's because it's empty," she said. "Mother says you'd feel better if you had something in it."

● Chrissy was getting tired of hearing her new baby sister scream. "Mom, didn't you say he just came from heaven?" she asked.

"That's right, Chrissy," the mother said with a smile.

"Well," the annoyed little girl remarked, "no wonder they threw her out."

● A farmer admiring babies in the maternity ward of the local hospital paused at one particularly cute infant. "Hmm, I wonder how old you are, baby," the man thought out loud.

Much to the farmer's surprise, the baby answered, "I'm two days old, sir."

"God in heaven!" the man gasped. "What is your name?"

"Jonathan Larry Frink," the baby replied.

"And where did you come from?" asked the farmer.

The baby very matter-of-factly explained the miracle of birth.

"Incredible!" the man exclaimed. "How do you know all this?"

"Sir, please," the baby said, "I wasn't born yesterday."

● A certain country doctor plays a game with some of his young patients to test their knowledge of body parts. One day, while pointing to a boy's ear, the doctor asked, "Is this your nose?"

Immediately, the child turned to his mother and said, "Mom! I think we better find a new doctor!"

● Little Pete was getting his first barber shop haircut. Once he was settled in the chair, the barber said to him, "Well, young man, how would you like you hair to be cut?"

Pete replied with a grin, "With a hole in the top—like Dad's."

● "Grandma, were you really once a little girl like me?" asked Winnie with her eyes open wide.

"Why, yes, dear," answered Grandma, finding the little one's innocence so charming.

"Well, then," Winnie remarked, "I suppose you know how it feels then to get a brand new doll when you don't expect it."

● A Boy Scout leader had a number of his group out on a camping exercise and was trying to teach them the fundamentals of outdoor cooking. After watching several new members at their attempts at this form of cooking, he asked, "How are you managing, fellows? Have you forgotten any essential equipment?"

"Yes, I have," said Billy, throwing down his tin pot.

"And what might that be?" asked the leader.

"My mother," the boy huffed.

It's Who You Know That Counts

The little girl was alarmed by the furiously barking dog. "Does he bite?" she asked nervously.

"Not if he knows you," assured the owner.

"Well, my name is Mary Elizabeth Clements," she said. "Tell your dog, will you, please?"

Chapter 23

Reading, Writing &'Rithmetic

"As long as kids have tests, there will continue to be prayers in school."

● A third grader was counseling his younger brother who was just getting ready to start school.

"Remember, in school, when the teacher says, 'No,' she means it—it's not like around here."

● "Now, Richie," said the country school teacher, "can you tell me what a hypocrite is?"

"Sure," Richie confidently replied. "It's a boy who comes to school with a smile on his face."

● A farmer was looking over his young son's report card. "Well, Stanley," the dad said, "with these grades, one thing is for sure. You couldn't possibly be cheating."

● "Mom," said Little Jake, "today our teacher asked me whether I had any younger siblings."

"Huh, I wonder why," said Jake's mom. "What did you tell her?"

"I told her I was an only child."

"And what did she say?"

Little Jake answered, "'Thank goodness!'"

● Miss Larkin asked the young farm boy, "Pete, how was it that your arithmetic homework was all correct for once?"

Pete replied, "My dad was out of town at a cattle auction."

● A teacher was giving a lesson on the importance of obedience. She proceeded to tell the story of an independent lamb that strayed from its flock, only to be eaten by a wolf.

"So, you see," the teacher explained, "if the lamb had obeyed and stayed with the flock, it wouldn't have been eaten by the wolf, now would it?"

"No, ma'am," one student replied. "It would have been eaten by people."

● A small farm boy came home with his report card, which left much to be desired. His father read it, frowned and before he could make a comment, the boy asked, "Well, what do you think is the problem—heredity or the environment?"

● The farmer asked his son what he had learned in school that day.

His son replied, "I learned to say 'yes, sir' and 'no, sir' and 'yes, ma'am' and 'no, ma'am'."

"Really?" his impressed father said.

"Yup," came the reply.

● The first-grade field trip kids saw a flock of birds about to migrate. The teacher explained that they were noisy and excited because they were starting out on a long journey. "What so you suppose they are saying?" she asked.

A shy little girl spoke up, "I bet the mother birds are telling all their children they'd better go to the bathroom first."

● Two young farm children were patiently waiting for the school bus one morning. Kenny observed, "Do you realize that after we finish all these years of school we'll have to find a job and then work for the rest of our lives?"

Completely taken aback, all Sally could say was, "And who came up with that idea?"

● One day, instead of serving the usual hot meal, the school cafeteria handed out peanut butter and jelly sandwiches.

After lunch, a satisfied first grader marching out the door complimented the cafeteria manager: "Finally, you gave us a homecooked meal!"

● Lenny sheepishly announced to his father, "There's going to be a small parent meeting tomorrow."

"Well, if it's going to be a small one, do I have to go?" asked the father.

"Ah, kind of," answered his son. "It's just you, me and the principal."

● Overheard in a teachers' lounge: "Not only is he the worst behaved kid in the class, but he has a perfect attendance record."

● Timmy came home from school with his report card—all zeros.

His father grumbled, "What's the matter—they run out of stars?"

Timmy smiled and said, "Uh, yeah, Dad, now they're giving out moons."

● The cook in an elementary school cafeteria served venison one day and asked the kids to guess what it was. The hint given to the children was, "It's what your mother sometimes calls your father."

At that, one little boy screamed to the rest of the class, "Don't eat it! It's jackass!"

● The rural schoolteacher was discussing the wonders of modern science, and in particular various kinds of machines. Asking the nine-year-olds what were the most wonderful machines they had ever seen, she got all the common answers—airplanes, television, tractors—until she called on one very thoughtful little girl.

"OK, Amy," the teacher said, "and what is the most wonderful machine you know?"

A slight pause, and then, very confidently, the girl answered, "A hen."

"A hen?" repeated the teacher. "Whatever makes you say that?"

"Well," replied Amy, "do you know anything else that will take all our scraps from the table and turn them into fresh eggs?"

● The math lesson topic today was subtraction, and the teacher formulated a word problem to illustrate the principle.

"Norman," said the teacher, "if you had seven pieces of candy and I asked for four, how many would you have left?"

Norman replied, quite honestly, "Seven."

● The teacher was circulating around the room while his class was taking a math test. He stopped at the desk of a boy who seemed to be having difficulty figuring out a problem.

Offering some assistance, the teacher asked, "How far are you from the correct answer?"

The student glanced around the room and said, "Uh, about two seats."

● On the first day of school in the small rural community, each kindergartner arrived home with a note from the teacher. It read...

"*Dear Parents:* If you promise not to believe all your child says happens at school, I'll promise not to believe all he says happens at home."

● Three little boys in the back of the room had their heads together, whispering.

"What's going on back there?" inquired the teacher.

"We're telling dirty stories," confessed one of the boys.

"Thank goodness," said the relieved teacher. "I was afraid you were praying which is against the law in school."

● The teacher had occasion to scold a small boy for swearing.

"If you must say something, just say 'darn,'" she said. "Your father doesn't swear, does he?"

"Oh no!" the boy said.

"Well, then, if he were working in the garden, and suddenly stepped on a rake which flew up and hit him in the face, what would he say?"

"He'd say, 'You're back early, dear!'"

● On the first day of kindergarten, a five-year-old boy was nervous and upset and wanted to talk to his mommy. A teacher helped his make a phone call to his mother, but when his mother answered the phone, he was too insecure to speak right away. So, the mother, on the other line, said, "Hello, who is this?"

"This is your son," said the little boy, bursting into tears. "Have you forgotten me already?"

● "With a single stroke of the brush," said the tour guide, leading the group of school children through the art museum, "Joshua Reynolds could change a smiling face into a frowning one."

"So can my mother," said one of the small boys.

● "Lucy! I hope I didn't see you looking at someone else's paper!" the teacher warned her student.

Lucy looked up at the frowning teacher and said, "I hope so, too!"

● Wife to husband, who is helping their small son with his homework: "Good, dear, help him now while you can. Next year he goes into fourth grade."

● The teacher had asked the class to list, in their opinion, the nine greatest Americans. After a while, she stopped at one desk and asked, "Have you finished your list yet, Freddie?"

"All but one," he replied. "I can't decide on the shortstop."

● As part of her Thanksgiving lesson, the teacher had asked her third grade class to write a composition on the things for which they are most grateful.

One little boy summed it up very neatly. "I am most thankful for my glasses," he wrote. "They keep the boys from hitting me and the girls from kissing me."

● "Tommy, your paper is exactly the same as Joey's," the teacher scolded. "Even the misspelled words are the same. How do you explain that?"

"Well," Tommy explained, holding up his writing utensil, "I used his pencil."

● The rural school teacher was giving a lesson on addition. "If I lay one egg on the table and two on the chair, how many will I have?" she asked her class.

Little Willie piped up, "Personally, I don't think you can do it."

● "The principal thinks I'm very responsible," said the kid to Dad.

"That's great son," replied the proud father.

The youngster explained, "Yeah, every time something goes wrong in school, he thinks I'm responsible."

● An iron worker calmly walked the beams high up above the street amid the nerve-racking noise of pneumatic hammers. When he came down, a man who had been watching him said, "I was amazed at your calmness up there. How do you happen to be working on a job like this?"

"Well," replied the worker, "I used to drive a school bus, but my nerves gave out."

● "How did you do in your exams, Tommy?" a 10-year-old farm boy was asked.

"Oh," he answered, "I did what George Washington did."

"What do you mean?" his father inquired suspiciously.

"I went down in history!" was the reply.

● "Your handwriting is atrocious!" Jimmy's mother scolded, looking over her son's school work.

"But, Ma," the 10-year-old argued, "I find it to be a good thing when people can't read my writing."

"Mister, name one good thing that comes from poor penmanship!" argued his mother.

"Well," Jimmy began, "you didn't say a word about all the mistakes in my spelling!"

● A small farm boy had to be excused during a grammar lesson. "Can I go to the bathroom?" he asked his teacher.

"Did you say can?" replied the woman loudly, taking the opportunity to reinforce the day's lesson.

"No, ma'am," the boy explained just as passionately, "I said bathroom!"

● An eight-year girl came home from school and told her father that every day when she started home, some boys grabbed her and kissed her.

"Why don't you try to run away from them?" suggested her parent.

"I did try that," the distressed girl answered, "but they wouldn't chase me."

● The teacher placed this sentence on the blackboard: "I didn't have no fun over the weekend."

"Now, how could I correct this?" she asked her fourth graders.

A brave voice from the back of the room called out, "Get yourself a date!"

● A little boy at school ran up to his teacher, sobbing bitterly. "What in the world is the matter Kenny?" the concerned teacher asked.

"I don't like coming to school, and I just found out that I have to stay here until I'm 18!" Kenny wailed.

"Yeah, well toughen up, kid," the teacher responded, "I have to stick around until I'm 65."

● During a science lesson on chemistry, the instructor asked one student, "This gas contains poison. What steps would you take if it should by chance escape?"

The student was quick with his reply, "Long and fast ones!"

● The superintendent was visiting one of his country schools when he became very annoyed at the noise made by students in one of the classrooms.

At last, unable to stand it any longer, he opened the door and burst upon the class. Seeing one boy taller than the others and talking a great deal, he seized him by the collar, removed him to the next room and stood him firmly in the corner.

"Now, you stand there and be quiet!" the superintendent commanded.

Ten minutes later, a small head appeared around the door, and a meek voice asked, "Please, sir, may we have our teacher back?"

● Two small farm boys were walking home from school when they saw a boy from their class. "There goes teacher's pet," Patrick whined.
"Yeah," Fred agreed, "if he said two and two were four, she'd say he was right."

● Little Paul boldly handed his report card to his parents.
"My, you sure seem confident, young man," his father commented. "What a nice surprise."
"Yeah, yeah, yeah," Paul responded. "Do me a favor. Just look this over and see if I can sue for defamation of character."

● "So, Jack, how do you like school?" the visiting relative asked the little boy.
Jack's answer was right to the point: "Closed."

● A little girl came home from her first day at school and her father asked, "Well, honey, what did you learn today?"
"Nothing much," the little girl said glumly, "I've got to go back tomorrow."

The Stork Is Coming!

Mrs. Jones was trying to explain to her two preschoolers the fact that they were soon to be joined by a baby brother or sister. She diplomatically explained the present location of the expected offspring and noted that its presence was the result of a seed.

The younger of the two boys, eyes wide open with wonder, exclaimed, "What kind of seed?"

To which the older and more sagacious of the two promptly replied, "Why it's a bean seed."

That one completely flabbergasted Mrs. Jones, and she asked her son what he meant.

"Come on, Mom," the boy said. "Haven't you ever heard of a human bean?"

Chapter 24

The Teen Years... Parents' Purgatory

"Children grow up so quickly. One day you see that your car's gas gauge is showing empty, and you realize they're teenagers."

● "Say," George said, admiring his buddy's recent purchase, "how many miles per gallon do you get on your new car?"

George had to stop and think. "Well, Jack," he replied, "I get about seven and my teenage son gets the other 12."

● A teenage country boy was always asking his dad if he could borrow the family car. Pushed to the limit, the dad asked his son why he thought the Almighty had given him two feet.

Without hesitation, the son replied, "One for the brake and one for the gas!"

● The father of a teenage girl was introduced to his daughter's boyfriend, who had a glass of milk in one hand and a huge slab of pie in the other.

"I'm glad to finally meet you," said the father. "I've been seeing you in our grocery budget for some time now."

● "I hear your son's on the high school football team," said the neighbor. "What position does he play?"

"I'm not sure," said the boy's father, "but if he plays football anything like he goes about his chores, I'd say he'd be a drawback."

● Two farmers got into a conversation about their trouble-prone children. "My teenage son is finally at peace with himself," said Hal.

"Hey, that's great," answered Cal.

"Yeah," replied Hal, "but he still fights with everyone else."

● "Excuse me, but you need to go to the back of the line," the cashier told the droopy-eyed teen.

Two minutes later, he returned to the cashier.

"What's the problem? I thought I told you to go to the back of the line," the clerk said, obviously annoyed.

"Show's how much you know," the teen answered back. "Someone else is already there."

● A farmer was teaching his daughter how to drive in town when the brakes on the pickup suddenly failed on a steep downhill grade.

"I can't stop," she shrilled. "What should I do?"

"Brace yourself," warned the father, "and try to hit something cheap!"

● The peak of mental activity must be between the ages of four and 17. At four they know all the questions; at 17, all the answers.

● Mrs. Hooper was horrified to see her daughter kissing a boy she had just met at the county fair. "Kissing a man you just met! You never saw me doing that!" Mrs. Hooper cried.

"Well, no," the daughter replied casually, "but I'll bet Grandma did."

● Joey and Doug were walking home from school. "What should we do tonight?" asked Joey.

"I know," said Doug, "let's flip a coin. If it comes up heads, we go to the movies. If it comes up tails, we go to the football game. And if it stands on end, we study for our history test."

● A father was out shopping for a new car. "Tell me, sir," the slick salesman asked, showing the man around the lot, "what do you want most out of your new car?"

The reply came quickly: "My teenage son."

● Debbie was furious with her brother. "I thought I told you not to tell Mom and Dad what time I got home last night!"

"I didn't," teased the brother. "I just said I was too busy getting breakfast to notice."

● Some teenagers have no hang-ups. Everything they own is on the floor.

● Mrs. Davis was talking with her neighbor about her husband's recent fall. "Does your husband know what caused his fainting spell?" asked the neighbor.

"Oh, yes," Mrs. Davis replied. "Our son asked for the keys to the garage and came out with the lawn mower."

● At the end of a driving lesson with his daughter, Homer sighed as he studied his nervous pupil clutching the wheel.

"We still have a few minutes to kill," he said gently. "Shall I show you how to fill in accident forms?"

● The farmer was lecturing his son on the value of money. "You know, son, when I was a boy, there was a depression and my father didn't make much money. I didn't have half the things you do."

"Well," replied the boy, "can I help it if my father is smarter than your father?"

● A father introduced his teenage daughter to the parents of his old high school buddy. Much to the embarrassment of the father, the girl remained silent throughout the whole introduction.

Dad knew how to explain her silence in terms any parent could understand: "She's not used to talking until she hears the phone ring."

PIG IRON

● Lenny and Dale sat down to lunch in the school cafeteria. Making small talk, Lenny asked, "Hey, Dale, have you read any good mystery books lately?"

"Actually," Dale answered, "I'm in the middle of one right now."

"What's the name of it?"

"Algebra II."

● Two neighbors were observing some odd behavior down the street. "Why is Bart Brown pacing up and down in front of his house like that?" asked the first neighbor.

"He's terribly worried about his teenage son," answered the second.

"Is the boy sick? What does he have?"

"The car."

● With obvious reluctance, a high school student handed his report card to his father, who studied the card, then signed it. The boy glanced at the signature, then asked, "Why did you sign with an X instead of your name?"

"Because," his father said, "with your grades, I don't think your teacher would believe you had a father who could read or write."

● You know a boy is growing up when he looks at a girl the way he used to look at a plate of chocolate chip cookies.

● Farmer Ned raced out of the house to see his son pull up the drive in the family car. The car had obviously been in an accident, as the fender was a crumbled mess.

"Great news, Dad!" the son said, trying desperately to find a silver lining. "You haven't been pouring those insurance payments down the drain after all!"

● A mountaineer took his teenage son to a high school to enroll him.

"My boy's after larnin'. What d'ya have?" he asked the teacher.

"Well, sir, we offer English, trigonometry, spelling, all sorts of classes," she replied.

"Well, give him some of that thar triggernometry—he's the worst shot in the family."

● The chemistry teacher was giving his class a verbal quiz. "What," he asked, "is the most outstanding result of the use of chemistry in the past 500 years?"

"Blondes!" came the quick reply.

• Kids used to think money grew on trees. Now they know that it comes from automatic teller machines.

• It's obvious that teenagers don't like parents burdened with money. They often relieve them of that very burden.

• A teen and his date were out for a joyride, when the girl expressed some concern over the driver's recklessness.

"I wish you'd drive more carefully!" Penelope told Roger. "I get nervous when you gun your motor and go roarin' around these blind curves on two wheels!"

"Aw, heck," Roger answered, "just do what I do—keep your eyes closed!"

• A farmer had a visitor who asked him, "So, what is your son going to be when he gets out of high school?"

The farmer sadly replied, "From the way it looks now, probably a very old man."

• The saleswoman watched as the teenager twirled in front of the store's mirror. "I love this dress!" bubbled the girl. "It's totally perfect! I'll take it!"

Then the young shopper paused thoughtfully. "Just in case my mother likes it, can I bring it back?"

• Kevin was borrowing his parents' car, but only under the condition that he drop his mother off at her friend's.

"It's amazing," Mrs. Davis commented to her friend as she walked up the front steps, "how fast the teenager who couldn't learn how to run the vacuum cleaner or lawn mower learns to drive the family car."

• The teller at the country bank's deposit window sharply reprimanded a man because he had neither filled out a deposit slip nor put his loose coins in special little rolls of specified amounts.

"When you've done this properly, I'll be glad to accept your deposit," the cranky teller said.

The man meekly accepted this tirade and went to a counter to follow instructions. When he returned to the window, the teller had since calmed down and had begun to apologize for his rude behavior.

"Oh, that's all right," the man said. "I have a house full of teenagers, so I'm used to being spoken to like an idiot."

The Well-Educated

Chapter 25

"Education is that wonderful tool that enables you to earn more than your educator."

● Having just graduated from college, a young man was given a job in his father's business. The first morning he reported for work his father instructed, "Here, go out and sweep the sidewalk in front of the store."

"But, Dad," the graduate protested, "I'm a college graduate!"

"Oh, right," said the father. "Just a minute. I'll come out and show you how."

● Mrs. Dowling bumped into an old friend at the market. "I hear you have a boy in college," the friend said. "Is he going to become a doctor, an engineer or a lawyer?"

"That I don't know," answered Mrs. Dowling. "At this point the big questions is: Is he going to become a sophomore?"

● "It's a terrible thing," grumbled Farmer Zeke. "I sold my car and mortgaged my farm just to send my son to college. And all he does is screw around and chase girls."

"You're regretting it, eh?" asked his sympathetic friend.

"You're darn tootin'!" snorted Zeke. "I should have gone myself!"

● Farmer Pete's son, away at an agricultural college, wrote the following letter home:

"Dear Dad, Gue$$ what I need mo$t of all. That'$ right. Plea$e $end it along. Be$t wihe. Your $on, $tephen."

But ol' Pete was not to be outdone. He wrote back:

"Dear Stephen, NOthing ever happens here. We kNOw you like your school. Write us aNOther letter. NOw I have to say good-bye. Your Dad."

● Farmer Roy asked his son if he was in the top of his class.

"Not exactly," he mumbled. "But I am one of those who make the top half possible!"

● Two farmers were griping about their college-aged sons home for the summer. "My son, the college man!" one complained. "Why, he can't even tell when the lawn needs mowin'!"

"Tell me about it," agreed the other. "I'm beginning to think college bred really means a four-year loaf made from the flavor of youth and his old man's dough."

● A photographer was taking a picture of a farmer and his college-student son. He suggested that the boy stand with his hand on his father's shoulder.

"I think it would be more appropriate," said the grumpy father, "if he stood with his hand in my pocket."

● A dairy professor, driving near a university campus, saw a student in an all-out sprint. Snarling at his heels were three huge dogs.

Intent on rescue and fueled by his own fear of large dogs, the professor screeched his Volkswagen to a halt and threw open the door. "Get in! Get in!" he commanded.

"Right on!" the bearded youth gasped as he approached the car. "Say, you're the greatest. Most people won't offer a ride when they see I have three dogs."

● An agricultural college professor, who suspected his class was dozing off on him, decided to see if anyone was paying attention. So, he suddenly continued his lecture in nonsense talk.

"You then take the loose sections of fendered smolg and gwelg them—being very careful not to overheat the broughtabs. Then extract and wampf them gently for about two minutes. Fwengle each one twice, then swiftly dip in blinger, if handy. Otherwise, discriminate the entire instrument in twetchels. Any questions?"

"Ah, yeah," came a sleepy voice from the rear. "Can you explain what twetchels are?"

● Two farmers were discussing the fact that they both had daughters away at college. "What does your girl plan to be when she graduates?" asked one.

"I'm not really sure," replied the other. "But judging from her letters home, she'll probably end up as a professional fund raiser."

● Money isn't everything, but it sure keeps you in touch with your kids.

● "I think our son is home from college, dear," Donna told her husband.

"What makes you think that?" he replied, keeping his nose buried in the newspaper.

"Well, we haven't had a letter from him asking for money in three weeks," Donna replied. "And the car is missing."

● "My son just graduated from agricultural college," the mother proclaimed.

"Did he win any honors?" her fiend asked.

"Yes," the mother replied sourly, "he was voted the most likely to sack seed."

● "When are you going to fix that fence, Larry?" nagged his wife.

"Oh, next week when Todd comes home from college," Larry replied sleepily.

"But what'll the boy know 'bout fixin' fences?" the wife yelled.

Larry replied smugly, "Well, he oughta know a heap. He wrote me that he'd been takin' fencin' lessons for a month."

● Farmer Lou was writing back his son who was away at college: "Dear son, I am enclosing $10 as you requested in your last letter. By the way, $10 is with one zero, not two."

● The confident young college grad walked briskly into the large bank. He stepped up to the manager's desk and began, "Good morning, sir. Has your bank any need for a highly intelligent, college educated man to make farm loans?"

The manager handed him an application form and offered to assist him in filling it out.

"Your name?"

"Daniel Pyles."

"Experience?"

"Just graduated from college."

"I see," said the manager. "And what kind of position are you seeking, son?"

"Well, I want something in the executive line, a vice presidency, for example," said the college grad.

The manager dropped his pencil. "I'm really very sorry," he said. "We already have 12."

The young man waved his hand and said confidently, "Oh, that's all right. I'm not superstitious."

● "For Christmas," Mrs. Jones remarked to her young daughter, "I was visited by a jolly, bearded, red-nosed fellow with a bag over his shoulder."

"Was it Santa Claus, Mommy?" the excited girl asked.

"No, dear," she replied, "it was your brother."

● A farmer, whose college son constantly complained about his lack of transport, finally agreed to finance some wheels, but only under one condition: The boy would get a haircut and shave off the beard he'd grown since he left the farm for school.

"I don't know why you're so opposed to beards, Dad," the son said. "After all, Jesus had long hair and a beard."

"That is true," said the farmer. "Jesus did have long hair and a beard. And he also walked everywhere He went."

- A farmer's son, attending the university to pursue a degree in music, complained to his landlord that the people in the upstairs apartment had annoyed him the previous night. "They were stomping and banging on the floor until well past midnight!" he exclaimed.

"So you couldn't sleep?" the landlord asked.

"Oh, I wasn't trying to get to sleep," the student said. "I was practicing my tuba."

- A college student home for the summer was arguing with his father about his current report card. "How could you have gotten four Fs and one D last semester?" scolded the father.

"I guess I spent too much time on one subject," his son replied.

SHEEP AT THE WHEEL

● The primary problem with the cost of education these days is how to save money for your children's college education while still paying for your own.

● Moe and Bob liked to talk about the woes that came along with having teenage sons.

"My kid is applying to difficult colleges," Moe stated with a frown.

"Why, Moe, that's something to be proud of," said Bob. "But why the long face?"

Moe said, "Well, they're easy to get into, but difficult to afford."

● The best neighbor is one whose kids are old enough to be away at college.

● Sometimes you have to keep a positive attitude despite overwhelming odds. A young college graduate wanted to work for a well-known farm cooperative. It was prestigious and high-paying, but also very difficult to get a foot in the door.

The grad presented himself for an interview and unabashedly extolled all of his virtues. He said, "I'm ambitious; I'm industrious; I'm a quick learner; I've got excellent college grades and recommendations and I'm willing to start at the bottom and work my way up."

The personnel manager said, "That's fine. But we have no entry level openings right now. Why don't you come back in about 10 years?"

The youngster replied, "Would morning or afternoon be better for you?"

● Two farm mothers ran into each other at the country store. They soon got to talking about their teenage daughters away at college.

"I think my daughter is majoring in communication," said Mary.

"Really?" replied Fanny.

"Yeah," answered Mary, "at least that's what it seems like from the phone bills she sends me."

● Why God Never Received Tenure At The University

— Because He had only one major publication.
— And it was in Hebrew.
— And it had no cited references.
— And it wasn't published in a journal or even submitted for peer review.
— And some even doubt He wrote it himself.
— It may be true that He created the world, but what has He done since?

— His cooperative efforts have been quite limited.
— The scientific community has had a very rough time trying to replicate His results.
— He never applied to the ethics board for permission to use human subjects.
— When one experiment went awry, He tried to cover it up by drowning his subjects.
— He rarely came to class, just told students to read the book.
— He expelled his first students for learning.
— Although there were only ten requirements, most students failed His tests.
— His office hours were infrequent and usually held on a mountain top.
— When subjects didn't behave as predicted, he often punished them, or just deleted them from the sample.

● In a college town, an old horse worked around the grounds for more than 20 years. The horse was a favorite for students and faculty who loved to visit with the animal.

After he died, the horse was given an honorary degree.

Incidentally, it was the first time a whole horse (not just his butt) had ever received this honor.

● Oliver Wendell Holmes once mistook an insane asylum for a college. Realizing his mistake, he joked to the security guard outside, "I suppose, after all, there is not a great deal of difference."

"Oh, yes, there is," replied the guard. "In this place, you have to show some improvement before you can get out."

Come Again?

● I used to be a lumberjack, but I couldn't hack it, so I got the ax.
● I used to be a doctor, but I ran out of patients.
● I used to work in a muffler factory, but I found the work too exhausting.
● I used to work in an orange juice factory, until they canned me. Yup, they put the squeeze on me so badly, I couldn't concentrate.
● I used to be a musician, but I wasn't noteworthy.

Chapter 26

The Twilight Of Our Lives

"You know you're getting old when the candles cost more than the cake."

—Bob Hope

● Carl had noticed recently that Trudy, his wife, didn't respond to questions he asked when he spoke to her from a distance. He was becoming worried that she might be losing her hearing.

He didn't know how to bring up such a delicate subject to his beloved wife, so Carl decided to ask the family doctor for some advice.

"Well," the doctor suggested, "try this. Tonight while Trudy's fixing supper, stand about 15 feet behind her and ask her what she's cooking. If she doesn't reply, do the same thing, but move five feet closer. Do this until she responds. Let me know what happens, and we'll figure out what to do from there."

That evening, while Trudy was at the sink washing lettuce, Carl stood about 15 feet away, cleared his throat and asked, "Trudy, what's for dinner?"

Carl got no response, so he moved up to about 10 feet away and tried again. Still, no reply.

He moved up again this time almost right behind her and asked, "Trudy, what's for dinner?"

Turning around to face her husband, Trudy groaned, "For heaven's sake, Carl! I've told you twice already! We're having chicken and biscuits."

● An old retired farmer had taken a room in a hotel and was preparing for bed. Just as he slipped under the covers and was reaching to turn out the light, the door opened unexpectedly and a radiant young blonde bounced into the room.

"Oh, excuse me!" she cried. "I must have the wrong room!"

The old man looked at her sadly. "Not only the wrong room, young lady, but you got here about 40 years too late."

● A tight-lipped spinster complained to the sheriff one summer about teenage boys swimming nude in a nearby stream, in plain view of her porch. The sheriff told the boys to move up the stream a bit. A few days later, the lady spoke to the sheriff again. "Haven't the kids moved?" he asked.

"They have," she snapped, "but if I go upstairs, I can still see them from the window." So the sheriff asked the boys to go still farther away. They said they would.

In a week, the lady was back in the sheriff's office. "They've gone upstream, all right," she said, "but I can still see them from the attic window through binoculars."

● An 83-year-old farmer was wandering through his swamp when he came upon a frog. The frog looked up at the farmer and said, "If you kiss me, I will turn into a beautiful 25-year-old woman that will marry you and satisfy you in every way imaginable."

The farmer scratched his chin, then picked up the frog and stuffed her in his pocket. The frog screamed, "What are you doing? Aren't you going to kiss me?"

"Nope," the farmer replied. "At my age, a talking frog is a lot more interesting."

● A friendly old farmer stopped in at the general store. While the owner was ringing up his purchases, the farmer said, "Skip, have I ever told you about my grandchildren?"

Skip quickly replied, "No, you haven't. And you don't know how much I appreciate it."

● Grandpa says, "You don't have to worry too much 'bout avoiding temptation after age 50. That's when it starts avoiding you."

● Jed and George were commiserating about the details that they have to tend to in their old age. "It's hard to be organized these days," said Jed. "There is sure a heck of a lot to take care of."

"Tell me about it," replied George. "I thought I was being organized when I prepaid my funeral expenses on my credit card. Then they asked for my expiration date. 'C'mon!' I said, 'I'm not that organized!'"

● An old farmer was sitting on a park bench talking to another old man. "I'm 90 years old and I have a beautiful young wife who loves me and caters to my every need. We live in a wonderful, multi-million house," he said as he started to cry.

"Well, why are you crying then?" asked his friend.

He said, "I can't remember where I live!"

● Back during the Great Depression, a farmer sent a check to Montgomery Ward for a case of toilet paper. A few days later, the check was returned along with a letter stating that the farmer would have to provide the catalog number of the item he desired. Without it, the order couldn't be processed.

The farmer was exasperated! "If I had the #%@#$% catalog, I wouldn't need the toilet paper!"

● The owner of a small travel agency looked out the window and saw an elderly couple gazing at the posters of exotic destinations.

He was immediately hit with inspiration: An all-expense paid vacation for the two of them. All he wanted in return was the endorsement of the senior travelers.

"That's all it is, folks," the travel agent said. "You both get the free trip and if you enjoy yourselves, you'll appear in TV commercials praising my travel agency."

The seniors looked at each other and agreed on the spot. "We'll take your offer," they said.

Two weeks later, they returned and the travel agent met them at the airport. When the man went to get the baggage, the travel agent took the woman off to one side and began asking questions.

"The food on the cruise ship was superb and the supersonic flight to Europe was a real thrill," she said. "The service was terrific, too."

"I'm glad to hear it," he said.

"But I do have one question about the trip," she said.

"I'll certainly try to answer it for you," he replied.

"Who was that old gentleman I had to sleep with every night?"

● A farm hand said to his manager, "Yesterday, Grandma fell down the stairs."

"Cellar?" said the manager.

"No," said the farm hand. "I think she can be repaired."

● "I've got very bad news. First, you've got cancer," the doctor told his elderly patient. "Second, I'm afraid you are going senile."

The patient replied, "Well, at least I don't have cancer."

● A small farm boy, doing his homework asked, "Grandpa, will you help me find the common denominator?"

"They haven't found that thing yet?" asked Grandpa. "They were looking for that when I was a boy in school!"

● An old timer decided to head into town for a movie. It was the first time he'd been to a movie theater in years.

He bought a box of popcorn and was aghast when the clerk told him it cost $3.50.

"My goodness," he exclaimed, "the last time I bought popcorn here it only cost 15 cents!"

"Well, sir," the attendant replied, "you're really going to enjoy yourself. Our movies have sound now."

● "There were exercise outfits in my day, too," Grandma said, "but then we called them house dresses."

● Two old farmers who had been rivals in their younger years bumped into each other at a community reunion. "Many of my friends who have known me for a long time say I don't look any different now than when I was 20 years old," bragged one of the old men.

"Is that so?" the other replied. "I can't imagine what it was like living over 50 years of your life looking like a 72 year old man."

● "I finally cured Pa of biting his nails!" proclaimed Old Mrs. Jones.

"Well, ain't that grand! How'd you do it?" asked her friend.

She said, "I just hid his false teeth!"

● Old Farmer Smith had just passed away, and his son was visited by the insurance agent.

"Tell me, son," the man said, "what was the cause of your father's death?"

The son had to think a moment. "Uh, I'm not sure," he replied. "But I don't think it was anything serious."

● Several elderly church members were asked to what they attributed their longevity.

"And why do you think God has allowed you to reach the age of 92?" one wealthy lady was asked.

Without hesitation, she answered, "I think He's testing the patience of my relatives."

● Mrs. Potts had struck up a conversation with Little Laurie. "I understand your grandmother is the old-fashioned type who still toils at the spinning wheel," she said. "Does she earn much?"

"Oh sure!" exclaimed Laurie. "Just last night she won $500 on red."

● "So, Henry," Farmer Ed said, "how's it feel to be a grandfather?" "Oh, it's nice to have a granddaughter all right," the new grandpa said with a smile. "Hardest part is getting used to the idea of being married to a grandmother!"

● The proud great-grandmother reported that the new baby had his father's nose and his mother's eyes.

"Yes," said the new father, keeping an eye on his father-in-law, "and if Great-Grandpa doesn't stop leaning over the crib, he's going to have his teeth as well."

● "Jimmy," said his mother one afternoon, "why don't you run across the street and see how Old Mrs. Weiss is?"

A few minutes later, Jimmy returned. "She says it's none of your business how old she is," he reported.

● Little Bobby took a long look at the old man and asked, "Were you on the Ark, Grandpa, when the flood came?"

"Of course not, Billy," replied his grandfather.

Next the boy asked, "Well, then why weren't you drowned?"

● Grandpa observes that a man is over the hill when he passes up a girlie calendar in favor of one that has room enough to write in all his grandchildren's birthdays.

● Old Farmer Vern was complaining to his friend over a game of chess.

"You know, Stanley, our generation never got a break. When we were young, they taught us to respect our elders. And now that we're older, they tell us to listen to the youth of our country!"

● Grandma and Grandpa couldn't help but notice the two teenagers kissing in the middle of the market's parking lot.

"Hmmm," Grandpa complained, "what's wrong with the younger generation?"

"The main thing, I think," commented Grandma, "is that too many of us don't belong to it any more."

● An elderly farmer complained to his doctor that he wasn't feeling well.

"I'm doing everything I can to help you," the doctor said. "You surely realize that I can't make you a young fellow again."

"I don't care about feeling young again," the farmer replied. "I just want to keep on getting older."

- Grandpa says, You know you're getting older when...
  - Everything hurts, and what doesn't hurt, doesn't work.
  - The gleam in your eye is the sun hitting your bifocals.
  - You feel like the night before and you haven't even been anywhere.
  - Your little black book contains only names ending in M.D.
  - You get winded playing cards.
  - Your children begin to look middle-aged.
  - A dripping faucet causes an uncontrollable bladder urge.
  - You know all the answers, but nobody asks you the questions.
  - You look forward to a dull evening.
  - You need glasses to find your glasses.
  - You turn out the lights for economic reasons rather than romantic ones.
  - You sit in a rocking chair and can't get it going.
  - Your knees buckle, but your belt won't.
  - Your back goes out more than you do.
  - You have too much room in the house and not enough in the medicine chest.
  - You sink your teeth into a steak, and they stay there.
  - You use tenderizer in your Cream of Wheat.

- An old farmer and his wife took a walk into town and began talking with a lady from the women's club. The woman asked the farmer if he had had wavy hair as a child.

"He has wavy hair now," laughed the farmer's wife. "Can't you see that it's waving goodbye?"

- The old farmer was in the hospital recovering from a serious operation. It was his birthday and he had received no cards or presents from his family.

The next day, three of his sons came to visit, all empty handed. After some brief conversation, the old man could stand it no longer and said, "Well, I see you all forgot it was your poor old Dad's birthday."

The sons were embarrassed and explained they had all been busy and it just plain slipped their minds.

"That's OK," said the old man. "I guess I can forgive you. Forgetfulness runs in our family. Heck, I even forgot to marry your mother."

"What?" exclaimed the sons. "Why that means we're..."

"That's right," replied the farmer. "And cheap ones at that!

Chapter 27

Keepin' Fit And Trim

"I joined a health club last year, spent $400. Haven't lost a pound. Apparently you have to show up."

—Rich Ceisler

● "The best thing for you, young man," the doctor advised, "is to give up drinking, smoking and women."

After some thought, the patient replied, "I don't deserve the best, Doc. What's second best?"

● Melvin was constantly being nagged by his wife to lose weight. Hoping to escape further criticism, Melvin liked to blame it on an overactive thyroid. One day, his wife dragged him to the doctor to put an end to his excuses, once and for all.

Much to his chagrin and the delight of his wife, the doctor made his diagnosis, "Your problem, Mel, isn't an overactive thyroid. It's an overactive fork."

● Two farm wives were chatting on the phone. "The town doctor told me I was carrying too much weight," confessed Mrs. Burns.

"Well, what do you think you can do about it?" asked Mrs. Franklin.

"Got the perfect solution," she answered. "From now on I'm having my groceries delivered!"

● Grandpa observes that about the only time over-weight makes you feel better is when you see it on someone you might have married.

● A young man was looking to his father for a bit of wisdom on staying healthy. "It says here if you study hard, don't drink or smoke or run around with girls, you'll live longer. Is that true?" the son asked.

"We don't know for sure, son," the dad answered truthfully. "Nobody's ever tried it."

● Farmer Dave's wife was reading a woman's magazine when she happened upon an article outlining a woman's "ideal" measurements.

"42-25-36," she said aloud. "I've just found out what's wrong with me. I'm upside down."

● Farmer Dugan was talking with this doctor.

"Do you think I'm too heavy?" the portly farmer asked.

"Well, Dugan," replied the doctor, checking his weight chart. "Let me put it this way. You're not overweight. You're just three feet too short."

● A girdle is a device used to keep an unfortunate condition from spreading.

● Farmer Irv and his wife were relaxing yet again in front of the TV, equipped with bowls of snacks.

"You know, Irv," the wife said, "the trouble with TV is that we like to spend so much of our time watching the 21-inch screen that we develop a 50-inch bottom."

● Wanda had just finished talking with her husband's ex-girlfriend at his high school reunion.

Walking away with her arm around her husband, she said to him, smiling, "Nothing is quite so gratifying to me as seeing a double chin on your old girlfriend."

● "My wife's arms must be getting shorter," Nelson announced.

"How's that?" replied his friend.

"Well, when we got married, she could reach all the way around me."

● "So, what made you finally decide to go on a diet?" Mrs. Henderson asked her husband as they cleared away the dinner dishes.

"Well, dear," he said, "I reckoned it was time to think about dieting when I tried to push myself away from the table, and it was the table that moved."

● Middle age is when the narrow waist and the broad mind change places.

● Two reasons why women don't wear last year's clothes: They don't want to, and they can't.

● A panhandler approached a rather large woman on the street.
"Lady, I haven't eaten in three days!" he cried.
"I wish I had your willpower," she replied unsympathetically.

● Gertrude and Helen were exchanging dieting tips.
"I heard that if you eat slowly, you will eat less," Gertie said.
"It's particularly true," replied Helen, the mother of six, "if you're a member of my family!"

● Grandpa observes that you've reached middle-age when it's a doctor and not a traffic cop who warns you to slow down.

● Sticker seen on a car bumper: The car is OK. It's the driver's body that needs work.

● When the doctor gave an overweight farmer a huge bottle of pills he told him with a grin, "Now, don't swallow these.
"Just spill 'em on the floor three times a day and pick 'em up one at a time."

● The farmer was sitting at the lunch counter busily scribbling figures on a piece of paper. The cafeteria owner became curious and asked him what he was doing.
"My wife is on a diet," replied the farmer, "and she told me she's losing four pounds a week."
"So?" queried the owner.
"So, if my figures are correct," explained the man, "I'll be completely rid of her in 19 months."

● The doctor asked the farmer to get on the scale and noted his weight. "You keep your weight fairly stable?" he asked.
"Sure do," answered the patient.
"What was the most you ever weighed?" asked the doc.
"About 185 pounds," said the farmer.
Doc added, "And the least?"
After thinking about his answer, the farmer replied, "Eight pounds, four ounces."

● "I really must start watching my waistline," the farmer said as he reached for yet another roll.
"Well, dear, how fortunate for you," teased his wife. "It's right out there where you can!"

● The overweight farm wife complained to her dieting companion, "This morning I called to sign up for an aerobics class, and the instructor told me to be sure to wear loose clothing.

"Can you believe it? If I had any loose clothing, I wouldn't need the class!"

● A fat farmer and a thin farmer collided on the street.

"From the looks of you, there's been a famine around here," the fat man jeered.

"And from the looks of you," replied the thin man, "you caused it!"

● A farmer was waiting for his wife after her aerobics class. As she came out of the room, he asked her if she had enjoyed herself.

The wife replied, "I'm not sure. With all that jumping, with all that pulling, with all that pushing... the class was almost over by the time I got my leotard on."

● Jerry's wife had been badgering him for months to take some "youth" pills she had been reading about.

Finally, he relented one night before going to bed and took several from a bottle she had bought for him. The next morning, his wife had to shake him to get him up.

"Wake up!' she stormed. "It's time you were out of bed."

"All right, all right," Jerry said defiantly, squinting one eye open, "I'll get up. But I don't wanna go to school!"

● Farmer Ned looked sad as he returned from an errand in town.

"What's the matter?" his wife asked him.

"I think it's about time I lost some weight," he admitted. "I got on the scale at the drugstore, put my quarter in and the card that came out of the slot said, 'One person at a time, please.'"

● Farmer Neal was conversing with his friend about his recent visit to the doctor. "My doctor was real subtle about my losing weight."

"Well, what did he say?" asked the friend.

Neal said, "He suggested I loan my body out to someone who'll exercise it."

● Two farm women were talking about the hardships of dieting. "The first week of a diet is pretty rough, but the second week isn't nearly as bad," said the first.

"Yeah, you're right," replied the second. "By the second week, I'm no longer on it."

● Gladys and Nancy were out shopping for the day and walked into their favorite clothing store. "You know," commented Gladys as she poked through the racks of slinky gowns, "I'll bet more diets start in dress shops rather than doctors' offices."

● "My wife has a new diet that allows her to drink anything that comes from the blender," remarked farmer John.

"How's it working out?" his neighbor asked.

"Great," said John. "Just last night, she drank two chickens, a cheeseburger and a pot roast."

● "I don't want to say my husband is fat," Stella told her friend, "but last night he told me he's planning on opening a chain of restaurants across the country."

"What does that have to do with his being fat?" the friend asked.

"Well," the wife replied, "he still hasn't decided if he'll open them to the public."

● Farmer Jake was returning to his doctor for his check up. "I bought myself a bicycle," Jake stated proudly.

"Do you use it much?" inquired his doctor.

"All the time," he replied. "I move it from one side of the garage to the other."

● A farmer was telling his friend about his wife's plans for dieting. "She bought a special book on reducing diets. It cost $12."

"How much has she lost?" his friend asked.

The farmer replied with a straight face, "Only $12."

● A newspaperman from an eastern city was traveling through a backwoods area when he happened upon a wrinkled, bent, old man rocking on his porch. Knowing that mountaineers are noted for their longevity and thinking there might be good story here, the reporter stopped to talk with the old man.

"Sir, I'd like to know your secret for long life," he said.

"Well," replied the old timer, "I drink a gallon of whiskey and smoke 25 cigars every day. I also go dancing every night."

"Remarkable," said the reporter, "and exactly how old are you?"

The mountaineer smiled a toothless grin and said proudly, "27."

● "You know your wife needs to lose weight," lamented Farmer Les, "when she wears two girdles—an upper and a lower."

● "My son finally decided to lose some weight," Mrs. Farmer announced.
"What helped him make that decision?" June replied.
"I think it was when he saw his graduation picture and noticed he not only was standing in the front row...he was the front row."

● "My neighbor's wife has got to lose weight," Farmer Fred told his buddy.
"Why do you say that?" the friend asked.
"Well, let's put it this way, she's two inches taller than her husband...when she's lying down."

● "Honey, you need to exercise more," Dorothy told her husband. "Why don't you try some stretching to loosen those muscles of yours?"
"If God wanted me to touch my toes," he replied, "he would have put them on my knees."

The Steak Diet

A woman asked a butcher, "Do you have a 15-pound roast?"
The butcher wrestled the roast out of the refrigerator and laid it on the counter. The woman looked at it, sighed happily and turned to leave.
The butcher, eager to make a sale, said, "It's U.S. Prime, lady. Doesn't it suit you?"
"Oh, I didn't want to buy it," the woman said. "I've just lost 15 pounds and just wanted to see what so much meat looks like."

Chapter 28

Is There A Doctor In The House?

"Three out of four doctors recommend another doctor."

● "I can't be sure what's wrong with you," the doctor said after examining his patient. "I think it's all the drinking."

"I understand," sighed the patient. "So, can we get an opinion from a doctor who's sober?"

● Farmer Rich was being up front with his physician. "Doctor, is there anything wrong with me? Don't frighten me half to death by giving it a long scientific name. Just tell me in plain English."

"Well, Rich," the doctor replied, "to be perfectly frank, you are just plain lazy."

"Thank you, Doctor," Rich said. "Now give me the scientific name for it so I can tell my family."

● A farmer was speaking frantically into the phone, "My wife is pregnant, and her contractions are already only two minutes apart!"

"Calm down, sir," the doctor said. "Is this her first child?"

"No, you idiot!" the man shouted. "This is her husband!"

● Despite repeated visits to the big-city hospital, doctors could not diagnose an old farmer's illness. Frustrated, the farmer returned to the team of doctors fully dressed from the changing room and asked, "Well? Will you be able to save me?"

The lead doctor scratched his head, then asked the patient to stand up. "Let me check one more thing," he said, moving his hands along the old farmer's hind quarters.

"I think so," he said encouragingly. "Come back tomorrow for another appointment." And the farmer departed with a big smile on his face.

As the medical team left the room, one of the medical students on the team stopped the lead doctor. "That was incredible, sir. My, what you did for that man's outlook on life! What did you check to verify whether or not this man could be cured?"

The seasoned doctor nonchalantly replied, "The size of his wallet."

● A surgeon was at a dinner party where the host was carving a roast. "See, Doc?" the host bragged. "I would definitely make a great surgeon, wouldn't I?"

"Anyone can take them apart," the surgeon replied. "Now put it back together."

● A surgeon was making hospital rounds and was chatting with a young lady he had operated on a few days earlier.

"Do you think the scar on my stomach will show?" she asked the doctor.

"That, my dear," he said, "is entirely up to you."

● "Dad, did you see the stork that brought me?" the youngster asked.

The farmer put his hand on his son's shoulder and said, shaking his head, "Only his bill, my child, only his bill."

● A medical student spent his summer vacation working as a butcher in the day and a hospital orderly at night. Both jobs, of course, involved wearing a white smock. One evening he was instructed to wheel a patient on a stretcher into surgery.

The patient looked up at the student and let out an unearthly scream. "My God!" she wailed. "It's my butcher!"

● "Say," said Roscoe, "did you ever see an eye doctor about that bad pain you get when drinking coffee?"

"Yeah," the farm hand replied. "He told me to remember to remove the spoon from the cup before drinking."

● The farmer went in to see his eye doctor one afternoon. "I always see spots before my eyes," he complained.

"Didn't the new glasses help?" the doctor said.

"Sure," the patient responded, "now I see the spots much clearer."

● "Doctor, Doctor," the farm hand pleaded, "you've got to help me. I just can't get my hands to stop shakin'!"

"Interesting," the doctor said. "Do you drink a lot?"

The worker replied, "Well, not really—I spill most of it."

● The farm daughter told her mom that a little boy in her class asked her to play doctor.

"Oh my," the mother nervously sighed. "What happened, sweetie?"

"Nothing," the girl said matter-of-factly. "He made me wait 45 minutes and then double-billed the insurance company."

● "The doctor said he would have me on my feet in two weeks," the farmer told his friend.

"And did he?" the friend asked.

The farmer said, "Yup, I had to sell the car to pay the bill."

● The town doctor was demonstrating the evils of alcohol to a drunk. He placed a worm in a glass of water, and nothing happened. Then he placed another worm in a glass of whiskey, where it shriveled and died.

"What does that tell you?" asked the doctor.

"Well," mumbled the drunk, "I'll never have to worry about worms."

● The farmer had been waiting in the outer office to see the doctor. Suddenly, an elderly woman charged out of the doctor's office and ran screaming down the hall.

"What's wrong?" the man asked nervously.

"I told her she was pregnant," replied the doctor.

"But surely that can't be the case, right?" asked the farmer.

"No," the clever physician replied, "but it sure cured her hiccups."

● The nervous patient said to the country doctor, "Doc, I need something to stir me up and get me in fighting trim. Did you put something like that in my prescription?"

"No need to," the doctor answered. "You'll find all that in the bill."

● A farmer, active in his rural community, went for a checkup for the first time in his life. When he presented himself without clothing, the doctor let out a screech. "My God! Your navel is 10 centimeters lower than normal. Was it always that way?"

"Um, no," the farmer confessed.

"Is it a work-related injury?" asked Doc.

With that, the concerned doctor rattled off a list of 25 questions, as the farmer shook his head to each.

Finally, the patient said to the perplexed physician, "Maybe it's because I've been the lodge's flag bearer for the last 20 years."

● The farmer was rushed to the hospital. He was in pain and could barely move. The doctor asked what had happened.

The farmer explained he had been stacking feed sacks in the barn. "The pile got too big and top-heavy and fell right on top of me," the farmer groaned.

"Well, from now, be more careful," the doctor said, shaking his head.

"See what happens when you don't practice safe sacks?"

● While examining a patient, the country doctor noticed the farmer had a scar on his scalp and asked him about it. "I got it from being drugged," the man told the doctor.

Failing to see the connection, the doctor asked him to explain.

"Well," the farmer said, "I was riding one of my horse. He bolted. My foot got caught in the stirrup. And I was drugged!"

● "My doctor finally agreed to make a house call to the farm," Willy declared.

"Wow!" Nels replied. "That doesn't happen often."

"Yeah," Willy said, "and just for fun, I kept him waiting in my living room for an hour."

● A waiter in a large restaurant was stricken and rushed to a nearby hospital's emergency room. On the operating table and in great pain, he waited for someone to help him. Looking around, the waiter recognized an intern who had recently eaten at the restaurant.

The patient pleaded with him. "Doc, I'm really sick. Can't you do something?"

"Sorry," said the intern as he hurried off, "this isn't my table."

● "Tell me the truth, Doc," the farm hand pleaded. "How long have I really got to live?"

"Well, son," the doctor said, "I don't want to alarm you, but from now on you pay me cash."

● Farmer Reid came home one day to surprise his family with a new car. "Dad," the son argued inspecting the vehicle, "this car ain't new."

"Well, all right," Bill confessed. "It's almost new. It was used by a doctor only to make house calls."

● When Mrs. Potter found her infant daughter eating handfuls of sand, she was frantic. She rushed the child into the house, made her drink lots of water and then phoned the doctor.

After explaining what had happened and what she had done, she asked Doc Turner what she should do next.

"Well, Mrs. Potter," Doc advised, "just don't give her any cement."

● A co-op manager was horrified to notice he was shrinking. Every day he was an inch shorter. In a panic, he went to see his doctor.

"I'm sorry," the receptionist said, "the doctor is very busy today, and he won't be able to see you."

"You don't understand!" the man wailed. "This is an emergency! I can't wait!"

"Now, sir," the nurse said without any emotion, "sit down and try to be a little patient."

● The reluctant farmer sat on the examining table at the doctor's office. "Well, here I am, Doc. I actually kept my appointment," he said.

"Good, now open wide," said Doc.

"Will that be my mouth or my wallet?" the farmer replied.

● "The doctor only gave me six months to live," Vern revealed. "So, I told him I wouldn't pay his bill."

"How'd he take that bit of news?" his friend asked.

"He gave me three more months!"

● A farmer had just come out of a six-week coma after falling off the roof of his farmhouse. "You've had a pretty close call, Mort," his doctor said. "It's only your strong constitution that pulled you through."

"Well," Mort replied, "remember that when you make out your bill."

● Farmer Stet was in the doctor's office yet again with another injury. "It's a bad sprain," the doctor told Stet, "but I'm not worried about it."

"Fine," the farmer replied, clutching his aching foot, "if your foot was sprained, I reckon I wouldn't worry about it either."

● Homer's wife ran into their family doctor at the country store. "Has your husband taken the medicine I prescribed?" inquired the doctor. "One tablet before each meal and a small brandy at night to stimulate his circulation?"

Homer's wife had to stop and think. "Well," she began, "He may be a few tablets behind, but he's months ahead on the brandy."

● The country doctor came to visit his patient in the recovery room. "Well, Judd, I've got some good news and some bad news," the doctor said.

"Give me the bad news first, Doc," Judd said nervously.

Doc said, "Uh, we amputated the wrong leg."

"Oh, my God! Please tell me the good news!" Judd cried.

"The good news," the doctor explained, "is that your other leg is getting better."

● A farm wife, a well-known hypochondriac, told her doctor in great alarm that she had a fatal liver disease. "Nonsense!" protested the doctor. "You wouldn't even know whether you had that or not. With that disease, there is no discomfort of any kind."

"I know," gasped the patient. "My symptoms exactly!"

● The new country doctor was very nervous as he opened his office doors for the first time. Self-conscious about how young and inexperienced he looked, the doctor prayed that he wouldn't say anything stupid to shake his credibility.

The first patient to sit on the examining table was a big, burly farmer, who looked very inconvenienced to have to pay the doc a visit for a few measly stitches. The doctor sensed the tension and attempted to make small talk with the disgruntled patient.

"So, how long have you been farming?" asked the doctor.

"Hmpf!" came the reply. "Doc, just put these stitches in so I can get back to my crops!"

Frazzled, the doctor said nothing more. Until, he happened upon a mark on the man's shoulder. "Say, what's that thing on your shoulder?"

"What?! It's a birthmark!" the grumpy farmer barked.

Stumbling for something to save him from this awkward moment, the doctor blurted out, "No kidding. How long have you had it?"

● A doctor had been sent an invitation to dinner. When the hostess received the reply, she found it completely illegible. "I have to know if he's coming or not," she told a friend.

"Why don't you take the note to a druggist," the friend suggested. "They're used to reading doctor's handwriting." Sounded like a good idea to the hostess, so she left for the local pharmacy.

The druggist looked at the paper and went into a back room. He returned shortly with a small green bottle. "Here you go," he said. "That will be $9.65."

● The surgeon told the farmer who woke up after having been operated on, "I'm afraid we're going to have to operate on you again," he said. "You see, I forgot my gloves inside you."

"Well," said the farmer, "knowing your fee, I'd rather just pay you for the gloves."

● A patient was meeting with his doctor about a diagnosis that would eventually lead to a very expensive surgery. "I don't know, Doc. The other physicians don't seem to agree with your diagnosis," the patient argued.

"That's fine," the doctor calmly replied. "The autopsy will show I was right."

- Things You Don't Want To Hear During Surgery...
  - — Oops!
  - — Has anyone seen my watch?
  - — That was some party last night. I don't remember ever being that drunk.
  - — &%$*#! Page 56 of the manual is missing!
  - — OK, now take the picture from this angle. This truly is a freak of nature.
  - — Better hold on to that. We'll need it for the autopsy.
  - — Come back with that, Spot!
  - — Hold on, if this is his appendix, what's this?
  - — Hand me that...uh...that uh...thingie.
  - — Has anyone ever survived 500 ml of this stuff?
  - — #%$@*!, there go the lights again.
  - — Ya know, there's big money in kidneys. Aw heck, this guy's got two of 'em.
  - — Everybody freeze! I just lost my contact!
  - — Could you stop that thing from beating? I'm having trouble concentrating.
  - — Sterile, schmeril. The floor's clean, right?
  - — What do you mean he wasn't scheduled for a sex change?
  - — This patient has already fathered some kids, right?
  - — Does anyone know if this guy signed his organ donation card?
  - — What do you mean "You want a divorce?!"
  - — Oh wow! Is that the fire alarm?

● "I have to pull this tooth," the dentist told his patient. "But don't worry, it will take only five minutes."

"How much will it cost?" the concerned farmer asked.

"About $90," the dentist replied.

"$90 for just a few minutes of work?" retorted the farmer.

The dentist answered smugly, "I can extract it very slowly for you if you'd like."

● A crop consultant walked into a doctor's office with a zucchini up his nose, a banana in his left ear and a carrot in his right ear. "Doc," he asked, "what's wrong with me?"

"It's obvious," the doctor replied. "You're not eating right."

● A farmer came into a dentist's office with his wife and said to the dentist, "Look, I want you to pull a bad tooth, but we haven't got a lot of time or money, so forget the Novocain."

"My, you are a man of courage," the dentist said. "Now, which is the bad tooth?"

The farmer turned to his wife and said, "Marie, show the man the tooth."

● Farmer Wes looked up at dimly at the white figure. "How 'bout it, Doc? Was my surgery successful?"

The reply: "I'm not your doctor, Wes. I'm Saint Peter."

● "The doctor says I've got to give up smoking," the farmer told his wife. "One lung is almost gone!"

"Can't you hold out just a little longer?" the wife responded. "I have almost enough coupons from the cigarette cartons for a new toaster."

● The overworked farm hand went to the doctor, complaining that when he touched his left leg with his finger, it hurt; when he touched his right leg, it hurt; when he touched his heart, it hurt; and when he touched his head, it hurt, too.

The doctor gave a quick diagnosis: "You must have broken a finger."

● Hank, a busy dairy farmer, remarked to his hired hand one day, "I finally went to the dentist yesterday."

The hired man replied, "Does your tooth still hurt?"

"I don't know," came Hank's reply. "The dentist kept it."

Higher And Going Higher

Grandpa was visiting the doctor for his annual physical.

"Gramps," the doctor began, "for a farmer of your age, you've got some funny ideas. I've been practicing for 40 years now, and I've never heard of a complaint like yours. What do you mean, 'your virility is too high?'"

Gramps responded with a sigh, "Doc, it's all in my head, and only in my head."

It's All In Your Head

Chapter 29

"Psychiatry is the care of the id by the odd."

● "Congratulations, Mr. Marshall," the psychiatrist told his patient, "you are completely cured of your delusions. But tell me, why are you so sad?"
Marshall replied, "Wouldn't you be sad if yesterday you were Leader of Planet Zoltar and the next day you were nobody?"

● The worried farmer called on his psychiatrist. "You gotta help me, Doc. All day long, I eat grapes."
"So what?" replied the doctor. "Lots of people eat grapes."
The farmer answered swiftly, "Off the wallpaper?"

● How many psychiatrists does it take to screw in a light bulb?
One, but it takes 12 visits.

● A farmer goes to his psychiatrist one day and said, "Doc, my brother thinks he's a chicken."
"Why don't you have him committed?" the doctor suggested.
"Are you kidding?" the farmer replied. "I need the eggs!"

● A farmer, feeling very stressed out over his strenuous work load and shaky marriage, took the advice of a friend and called a psychiatric hotline. The main menu recording was enough to send the poor guy over the edge...

— "If you are obsessive-compulsive, please press 1 repeatedly.
— If you are co-dependent, please ask someone to press 2.
— If you have multiple personalities, please press 3, 4, 5 and 6.
— If you are paranoid-delusional, we know who you are and what you want. Just stay on the line so we can trace your call.
— If you are schizophrenic, listen carefully and a little voice will tell you what number to press.
— If you are manic depressive, it doesn't matter which number you press, no one will answer."

● A pigeon flew over the rural mental hospital grounds and dropped a deposit on one of the park benches. A nurse jumped up and said she'd be back with some toilet paper.

"And they think we're nuts," said one inmate to another. "Why, by the time she gets back, that pigeon will be long gone."

● "The last time I met you," said the counselor, shaking her head, "you made me very happy because you were sober. Today you have made me very unhappy because you are intoxicated."

"True," replied the old drunk with a toothy smile, "but today's my turn to be happy."

● Two farm wives were gossiping about the widower down the street. "I heard he has schizophrenia," one said.

"Really?" replied the other. "Good for him."

"What do you mean 'Good for him?'" the first exclaimed.

The other wife answered, "Well, at least he doesn't have to eat alone."

● A psychiatrist had to call a plumber to fix a pipe that had burst. The plumber arrived, unpacked his tools, clanked around on the pipes for a while and handed the doctor a bill for $600.

HORSE DRAWN

"$600!" the doctor exclaimed. "This is outrageous! I don't even make that much as a psychiatrist!"

The plumber quietly responded, "Neither did I when I was a psychiatrist."

● "Now, tell me why you feel your parents rejected you," the shrink told his patient.

"Well, for one thing, there were those times when I would come home from school, and they weren't home!" said the man.

"Did it ever occur to you that they might be out for a walk or running errands?" the doctor asked.

The patient replied, "Yeah, but nobody takes the furniture with them when they run to the store."

● "Lie down on the couch," the woman's psychiatrist said.

"I'd rather not," replied the patient. "That's how all my trouble started!"

● The farmer told his shrink, "I'm always forgetting things—what should I do?"

The doctor replied, "Pay me in advance."

● A psychiatrist was telling his colleague about a patient who believed in voodoo and black magic. "He just doesn't realize that all that mumbo jumbo stuff is ridiculous. Voodoo is just a lot of superstition."

"You told him that, right?" the colleague replied.

"Are you kidding? And risk being cursed?"

● "I keep waking myself up with my snoring," the distressed patient lamented. "What can I do?"

The psychiatrist, filling out his bill, said simply, "Just sleep in another room."

● An overworked farmer dressed as Napoleon Bonaparte went to see the town psychiatrist per his wife's request. "So," the doctor asked, "what seems to be the problem?"

"Well, I have no problem," replied the farmer. "I am Napoleon Bonaparte, one of the most powerful people in the world. I have a great and strong army, all the money I will ever need and live in the lap of luxury."

"Then why did you come here?" asked the psychiatrist.

"I needed to talk to you about my wife," the patient remarked. "She think she's someone named Mrs. Jones."

Chapter 30

Stall-Side Manner

"Veterinarians: Doctors who never have to hear their patients' complaints."

• "Well, Mr. Turner, I have your coon hound's test results," the veterinarian explained. "But you might want to get a second opinion."
"Thanks, Doc," replied Turner, "but I might just be particularly happy with the first one."

• Vic the veterinarian was checking on Earl's prize heifer who had been suffering from some sort of stomach ailment. "Is there any hope, Doctor?" Earl anxiously asked.
"Well," the vet responded, rubbing his temple, "depends on what you're hoping for."

• A farmer was speeding home one day from a no-till conference when he swerved to avoid a deer and crashed into a tree. Fortunately, a passerby dragged him out of the smashed car and rushed him to the doc's office in the small town nearby.
"Afraid I'm the wrong fellow to help you," the doctor said. "I'm a veterinarian."
"Go to work on me anyhow, Doc," groaned the victim. "I was a jackass to go 60 on those tires!"

• The veterinarian said to the farmer, "Your horse needs an operation."
"I think I'd like a second opinion," the worried farmer answered.
"OK," the doctor offered, "he doesn't need an operation."

• A prospective buyer ran into a rancher/vet at a farm show. He asked him, "Hey, where's that horse you were going to sell me?"
"Oh," replied the man, "he got better."

• Two country vets were exchanging professional stories when one asked, "Have you ever made any bad professional mistakes?"
"Yes," the other confessed, "I once cured a very wealthy rancher's Thoroughbred horse in only two office calls."

● The ardent animal lover was most distressed because he ran over a rabbit and saw it lying in the road, drawing what appeared to be its last gasps.

He had just made up his mind to put the animal out of its misery when another motorist, a veterinarian, stopped, pulled out a bottle of tonic and placed it under the nostrils of the dying rabbit. In a few seconds, the animal got up and bolted through the brush. The animal lover was pleasantly astounded...and curious.

"What kind of stuff is that?" he asked.

"Hare restorer," came the sly response.

● A farmer's best sheep dog suddenly became very ill, so his wife drove the dog to the local veterinarian.

The vet came in and looked over the dog. He said to the woman, "He's barely moving. But there's one more test I can do to tell if he's still alive."

He then left the room and returned holding a cat. He carefully held the cat over the dog's tail, legs, body and finally his head. There was no response from the dog. The vet told the woman, "I'm sorry. Your dog is dead."

Two weeks later, she received her bill—an astonishing $1,000. She quickly sent the vet a note: "I don't understand this bill. You charge $1,000 just to tell me that my dog is dead?"

Three days later, a reply arrived from the vet. "Yes," he wrote, "it comes to $50 for the office visit and $950 for the cat scan."

● The struggling farmer was discussing the results of his ailing horse's check-up with his doctor. "Could you afford an operation if I found one necessary?" asked the vet.

The farmer scratched his head and said, "Ah, would you find one necessary if I couldn't afford it?"

● Clint's prize heifer had taken ill and the only veterinarian available to make a house call was the new one in town. The young vet went to the barn, but in a few minutes, came back to the farmhouse. "Do you have a cork screw?" he asked Clint.

The young vet took the cork screw and ran back to the barn. But several minutes later, he was back for a second time. "Got a screwdriver?" he asked. And away he bounded with the screwdriver in his hand. Almost immediately, he was back at the house again, asking for a chisel and a hammer...quickly.

The distraught farmer could stand it no longer. "For heaven's sake, Doc," he begged, "what's wrong with my heifer?"

"Don't know yet," came the hurried reply. "Can't seem to get my bag open."

● A veterinarian returned empty-handed and angry after trying to collect a long overdue bill.

"What's the matter?" asked his wife, "didn't you get your money?"

"No, I didn't," he grumbled. "And not only that, but the dog I treated bit me."

● In the middle of a blinding blizzard, the local veterinarian received a call from a farmer whose cow needed medical attention. The farmer pleaded with the vet to drive out to his farm. "I'd be glad to come, but my car is in the garage for repairs," the doctor explained. "You'll have to come and get me."

"What!" exploded the caller. "In this weather?"

● A vet sent a floral arrangement to a former colleague who was leaving his current position and opening his own animal hospital. Upon visiting his friend in his new place, he was horrified to see the florist had mistakenly sent over a wreath with a "Rest in Peace" banner on it.

Furious, the vet called the florist to complain. "It could be worse," said the florist. "Somewhere in this town there's a bouquet in a cemetery with a note that says, 'Good Luck in Your New Location.'"

● "I've got good news and bad news," the vet told Milo. "Which do you want to hear first?"

"Give me the bad news," Milo said, bracing himself.

"Your heifer's heart is about to give out; she doesn't have much longer."

"Oh God! And what's the good news?" Milo asked.

The vet said, "I broke eighty on the golf course yesterday!"

● A country veterinarian was feeling sick and went to see her physician. Her doctor asked her all the usual questions about symptoms, how long they had been occurring, etc., when she interrupted him. "Hey, look," she complained. "I'm a vet, and I don't need to ask my patients all these questions. I can tell what's wrong with them just by looking at them. Why can't you?"

The doctor nodded, looked her up and down, wrote out a prescription and handed it to her. "There you are," he said. "Of course, if that doesn't work, we'll have to put you down."

Leaders Of The Land

Chapter 31

"98% of the adults in this country are decent, hard-working Americans. It's the other lousy 2% that gets all the publicity. But then, we elected them."

—Lily Tomlin.

● "There are just so many candidates this year," commented Farmer Jack as he flipped through his election pamphlets.

"Yeah," agreed farmer Bill. "There may not be enough promises to go around."

● A very pompous Congressman was arrested out of his district for speeding. Unfortunately for him, he was brought before a judge just as proud as he.

Strutting before the bench, the politician spoke in his best Shakespeare-an manner, "I'll admit I may have been speeding a little, Your Honor, but you see, I'm a Congressman and..."

"Ignorance is no excuse," interrupted the austere judge.

● The candidate was to be a guest of honor at a banquet. As he finished dressing, he patted his cheeks, assumed a stiff posture and said to his wife, "I wonder how many really great men there are in this country today."

His wife, looking her husband up and down, said, "My darling, I don't know, but there is certainly one less than you think."

● A man running for office in the South hired two research assistants—one to dig up the facts and the other to bury them.

● "Whatcha doin', Jake?" the farmer asked his friend.

"I'm writing to our Congressman, regarding his ignorance about our local problems," replied Jake sternly. "By the way, who is our Congress-man, anyway?"

● A candidate, beginning his spiel in front of a group of farmers, advised, "My job, as I understand it, is to talk to you. Yours, from what I can tell, is to listen.

"If you finish before I do, just hold up your hand."

● An elderly farmer who had worked very hard for the election of a new state senator was very much surprised to find himself brought into court.

"What have I been arrested for?" he inquired.

"You are charged," explained the judge, "with voting seven times."

"Charged!" exclaimed the farmer. "I thought I was getting paid!"

● The emcee at a debate had this to say to the crowd of rural photographers covering the event for the town weekly: "Please do not photograph the speakers when they are addressing the audience.

"Shoot them just before they begin to talk."

● Two opponents for the same seat in the legislature met on a street corner and soon became engaged in an impromptu debate of their own. They soon drew a large crowd of spectators.

"There are hundreds of ways to make money," declared one of the candidates, "but only one honest way."

"And what's that?" inquired the other candidate skeptically.

"A-ha," cried his opponent. "I figured you wouldn't know."

● "Well, I reckon I can take this thing off my car," the farmer said as he peeled off the politician's campaign propaganda.

"Interesting, isn't it?" remarked his friend. "That guy's bumper sticker lasted a lot longer than his promises!"

● A candidate was talking to a member of the audience after his rather long speech. "And how did you like my talk, Farmer Jones?"

"Well, I found it very refreshing, really refreshing," Jones replied.

"Did you really?" asked the delighted speaker.

"Oh, absolutely," he said. "I felt like a new man when I woke up."

● A lot of Congressmen are unhappy because they're underpaid, underappreciated and under investigation.

● A young Congressman at the Capitol was an ambitious worker, and he had a very nice well-furnished office. However, he began to act very peculiar. First, he shoved his desk out into the space also occupied by his secretary's desk. Then a few days later, as he was heading out for the day, he pushed his desk out into one of the many long hallways. He worked there for a few days and then shoved his desk into the men's room and set up work there.

All of this was noticed by his coworkers. It seemed more and more strange to them, so strange that they did not dare ask the Congressman what he was doing. Instead, they went to the Congressional psychiatrist and asked him to find out.

So, the doctor walked into the men's room, sat on the edge of the Congressman's desk, and asked, "Why have you kept moving your desk? Especially, why into the bathroom?"

"Well," he said, "I figure that this is the only place in the Capitol where they know what they're doing."

● Two farmers, while waiting for their daughters to say good night to friends at the town dance hall, observed a group of teenagers dancing rather oddly. The teenagers would take two steps forward, one step back—and mix it up with a whole lot of side stepping.

"What the heck is that new dance called?" asked another farmer.

"I've never seen it before," replied the first farmer, "but after watching their moves, I reckon it should be called 'The Politician.'"

● Two retired farmers were shooting the breeze at the town pool hall. "There are two sure things in life: death and taxes," said Lyle.

"Yeah," said Jack, "but death doesn't get worse with every session of Congress."

● You know it's cold out when politicians walk around with their hands in their own pockets!

● Bill Clinton was walking through an exotic pet store when he noticed a small drab bird siting on a lonely perch. The bird looked at him and said, "Who are you?"

"Who am I?" echoed Clinton. "I'm the most powerful man in the world! I'm the president of the greatest country on this planet. Millions of people turn out just to see me walk down the street. My life is an inspiration to millions of young children! Now, little bird, who are you?"

"Me?" lied the little bird. "I'm an eagle."

● Why don't politicians play hide and seek?

No one would look for them.

● Grandpa observes that the cheapest way to have your family tree traced is to run for public office.

● The candidate had been talking on and on for almost an hour. Finally, he said, "Now, are there any questions?"

"Yes," came a voice from the back, "who else is running?"

● Two farmers were talking politics over the fence they shared. "I am opposed to limited terms for Congressmen," said Herman.

"Yeah," agreed Austin, "I say they should serve their full sentences."

● No wonder the people in government want to fix up our jails—look how many are winding up there.

● Dennis was reading the evening paper. "You know, Pam, those politicians used to kiss babies to show their emotional ties to the general public," Dennis said to his wife. "Now they wait until the babies are grown."

● Only in America do we used the word "politics" to describe the process so well: Poli in Latin meaning many , and tics meaning blood-sucking creatures.

● "To save money," said the chairman, "we've canceled dinner tonight. We figured you would rather hear the Senator speak than eat."

"Sure would," called a voice from the crowd. "I've heard him eat."

● Three brothers were sitting in a waiting room for an interview with the folks from the *Guiness Book of World Records.*

The first one, who claimed to have the world's smallest hands, went in for his meeting. About 20 minutes later, he returned jubilant—for the Guiness committee verified his record small hands.

HORSE CHESSNUTS

The committee called in the second brother, who claimed he had the world's smallest feet. He was in the office for the same 20 minutes or so and also returned excited, having made it into the *Book of Records.*

The third brother was now very anxious to describe his certainty that he possessed the world's smallest brain, and actually skipped into the office. His meeting, however, was very brief. He dejectedly returned to the waiting room where his brothers sat.

"What happened?" one asked.

He replied, "I was told there are more than 500 people in Washington D.C. already ahead of me."

● Four fellows were getting civic medals for their contributions to the community. The first, a doctor, announced, "I shouldn't get this medal. I have a confirmed weakness. After listening to patients' woes all day, I go home and take it out on my wife."

The second, a farmer, said sadly, "I shouldn't get it either. After all the problems I have on my farm, I go out each weekend and get stone drunk."

The third man, a police chief, with his head low, stated, "I'm certainly not worthy. I've been married for 20 years, but I can't stop running around with other women."

The last person, the mayor, said with a smile, "I'm sure I shouldn't get this award either. My weakness is gossip, and I just can't wait to get out of here!"

● Politics is the art of looking for trouble, finding it everywhere, diagnosing it incorrectly and applying the wrong remedies.

● Your sins may all be forgiven and never trouble you again—unless you decide to run for public office.

● While campaigning in a farm county, a politician ran into an unfriendly crowd.

Halfway through his speech, he was suddenly pelted with overripe fruit. His presence of mind, however, did not fail him. His next remark, as he wiped the mess from his face and shirt, turned boos into cheers.

"My critics," he said jauntily, "may not think I know about farm problems, but they'll have to admit I am a big help with the farm surplus."

● Two farmers were talking politics and lightened things up a bit by cracking jokes about the government.

"Ah, there sure is nothing wrong with a good political joke," laughed one farmer.

"Unless it gets elected!" joked the other.

● I guess the folks in Congress know what they're doing at least some of the time. You have to admit, it's pretty smart of them to put a high tax on liquor and then raise other taxes that drive people to drink.

● Two farmers were listening to an old politician who had unsuccessfully tried get elected for nearly a decade.

"I always feel optimistic after listening to this guy," one announced to his friend.

"Yeah, I know what you mean," he replied. "The next speaker can't possibly be as bad."

● A farmer was out working in his field one day when a carload of politicians raced by. They were going too fast for the curve and crashed. Soon afterward, the sheriff stopped by and asked the farmer if he had seen the car.

"Yup," replied the farmer, pointing to the ditch filled with fresh dirt.

"You buried them?" asked the sheriff. "They were all dead?"

"Two of 'em said they weren't," shrugged the farmer, "but you know how them politicians can lie."

When You're In A Jam...

To save everyone's time, give your excuse by number:

1. **That's the way we've always done it.**
2. **I didn't know you were in a hurry for it.**
3. **That's not my department.**
4. **No one told me to go ahead.**
5. **I'm waiting for an OK.**
6. **That's his job—not mine!**
7. **Wait till the boss comes and ask him.**
8. **I forgot.**
9. **I didn't think it was very important.**
10. **I'm so busy, I just can't get around to it.**
11. **I thought I told you.**
12. **I wasn't hired to do that!**

Stars And Stripes...

"The main object of the army is to promote the general's welfare."

—James C. Humes

● The new recruit from an indulgent home turned up his nose at the Army stew and complained to the mess sergeant, "Yuck! Don't I have a choice here?"

"Certainly, my boy," replied the sergeant, scooping up the drippy grub and slopping it on his tray. "Take it or leave it!"

● A small farm boy was leading a donkey passed an army camp. A couple of soldiers wanted to have some fun with the lad. "Say, little boy, what are you holding on to your brother so tight for?" laughed out one of the enlisted men.

The youngster stopped in his tracks, looked the soldier squarely in the eyes and said, "Well, I don't want him running off and joinin' the army, now do I?"

● The phone rang at the motor pool office of an army base and was answered by one of the many young privates assigned to duty there. "What kinda equipment you guys got down there?" asked the caller.

"Well, we got lots of stuff," came the reply. "We've got some big trucks for haulin' equipment around, we got some personnel carriers for haulin' people around and we got a Jeep for haulin' that half-witted colonel around."

"Just a minute, young man," interrupted the caller. "Do you know who you're talking to?"

"No, sir," answered the private.

"Well, this is Colonel Smith!"

"Do you know who you're talking to?" asked the private.

"No," replied the colonel.

"Well," came the answer, "so long, half-wit!"

● An Air Force captain and his Army son, a young man who had just finished his basic training, were traveling home on leave together. The father, in his dress blues, picked up his son, wearing his dress greens. They stopped along the way at a roadside diner for lunch.

As the waitresses brought their orders, she stared at them with a puzzled expression.

"Is there something wrong, ma'am?" the son asked.

"It's just unusual to see men in different services, especially an officer and an enlisted man, traveling together," she said.

The father winked at her and grinned. "Heck, that's nothing," he said. "This young man is taking me home to sleep with his mother."

● A group of young Navy recruits were undergoing a course in combat swimming. The program included jumping into a pool from a 12-foot diving board.

One of the recruits walked tentatively out to the end of the board, but he froze there and could not jump.

"You had better jump, boy!" his drill sergeant ordered.

Still, the fellow hesitated.

"What would you do," the drill instructor yelled, "if that diving board were a sinking ship?"

"Sir," the unnerved recruit yelled back, "I'd wait until the ship sank about 10 more feet!"

● A certain Navy captain and his chief engineer argued as to which of them was the more important to the ship. Failing to agree, they resorted to the unique plan of swapping places.

The engineer ascended to the bridge, and the captain went into the engine room. After a couple of hours, the captain suddenly appeared on the deck covered with oil and soot.

"Chief!" he yelled, wildly waving aloft a monkey wrench. "You have to come down here; I can't make 'er go!"

"Of course you can't," replied the engineer. "We're aground!"

● An Army recruiter drove into a small farm town. Everywhere he looked, he saw targets with bullet holes right smack in the middle.

Determined to find the "sure shot," he went into the local tavern and asked the barkeep if he knew about this. The old bartender smiled and pointed at the town drunk slumped over in the corner.

The Army recruiter walked over, struck up a conversation with the drunk and asked him how he had such accuracy when he was inebriated most of the time.

The drunk responded, "Aw shucks, sir, I just shoot first and draw the targets later."

Troublesome Taxes...

"A fool and his money are soon parted—the rest of us wait to be taxed."

● A farmer went into a restaurant and said, "Give me the tallest glass you've got and a lemon."

The farmer squeezed the lemon and got nearly a pint of juice. Tossing aside the lemon rind, he said, "I'd like to see anyone else get that much juice out of a lemon."

A little fellow, standing nearby, announced, "Let me have a tall glass and the lemon rind you just threw away." He then took the lemon rind and squeezed another full glass of juice from the already used rind.

"Man!" the farmer exclaimed, "I never saw anything like that! How did you do it?"

"Well, you see," the man replied quietly, "I'm with the Internal Revenue Service."

● Farmer Hank was seeking tax advice from a farmer friend.

"I need an accountant who will save me the most money possible," Hank told his friend.

"Listen, why don't you check into my accountant," the farmer suggested. "He came up with so many extra deductions for me last year that I was even able to post bail for him."

● The other day, a farmer got some bad news from the IRS and was telling his buddy about it. "They sent me a letter telling me I can't write my neighbor's kids off as dependents!" Jerry complained.

Bob was confused. "Why would you think you could write off your neighbor's kids?"

"Hey!" Jerry retorted, "If they can eat your food, ruin your lawn and make your wife furious...they should be deductible!"

● During his job interview for the IRS, the farmer's son was asked, "So, what can you do?"

He shrugged his shoulders and said, "Nothing."

The interviewer replied, "I'm sorry, sir, we have no executive-level openings at this time."

● The Internal Revenue Service is the closest thing to a Chinese dinner; no matter how much you gave them, a year later, they're hungry again.

● "Maybe I should just get myself a book about how to figure out my taxes on my own," said a disgruntled farmer to his buddy.

"Sure," his friend replied, "there are dozens of 'em out there. Only problem is, none of 'em has a happy ending."

● The income tax guys must love poor people—they're creating so many of them.

● A farmer received a second notice that the tax payment on his farm was overdue. The next day he went to the township hall, made out a check and apologized for overlooking the first notice.

"I'll tell you a little secret," said the tax collector with a smirk. "We don't send out first notices. We've found second notices are so much more effective."

● Taxation without representation was tyranny, but it was a lot cheaper.

● Farmer Zeke and his wife were busy at the kitchen table, working on their taxes.

"You shouldn't be grumpy about taxes," said his wife. "Just think how wonderful it is to live in a country like ours. And how lucky we are! You should pay your taxes with a smile."

"Well, I'll be glad to, Ma," Zeke answered, "but do you think the government would go for that?"

● A silver lining: There is no child so bad that he can't be used as an income tax deduction.

● The new accountant was trying to rally up business around tax time. "I promise," the accountant pledged, "if the tax returns I file for you ever get audited...I will feed your dog until you get paroled."

● A farmer called the IRS to see if he could get a certain deduction on his taxes. The reply was a hearty "No!" followed by, "This is a recording."

Chapter 34

Law and Order

"We ought never to do wrong when people are looking."

—Mark Twain

- Two Wisconsin policemen found a dead body on one town's odd-named street, Kinnickinnic Avenue. As one of the officers began filling out the report, he asked his partner, "How do you spell Kinnickinnic?"

"K-I-N...," the other cop responded slowly. "K-I-N-N...Oh, what the hell, let's drag him around the corner to Main Street."

- "Hey, did you hear about the nudist camp that was in the news last night?" the farm hand asked his co-worker.

"Can't say I did," he replied.

"Yeah, someone knocked a hole in the camp's wall," said the first worker. "Sources say, the police are lookin' into it."

- The loan officer at the rural bank had made a trip to the big city for a seminar on new lending procedures. As he was walking back to his hotel, he was jumped by two muggers. He defended himself valiantly, but they eventually subdued him.

His attackers then rifled through his pockets, but found only spare change.

"You mean you fought like that for only 57 cents?" asked one of them incredulously.

"Is that all you were after?" the loan officer moaned. "I thought you wanted the $400 in my shoe."

- While a sheriff was taking a prisoner to jail, a gust of wind blew off the prisoner's hat. The prisoner lunged for it, but the sheriff declared, "Oh, no you don't, wise guy. You just stand right where you are and I'll get it."

- An ill-tempered orchestra leader threw a tantrum, killing his principal violinist. He was sentenced to die in the electric chair, but when they threw the switch, nothing happened.

The man shook his head sadly and said, "I guess they were right. I am a bad conductor."

● After the small town bank was robbed for the third time straight by Wylie, a shifty character who managed to elude the sheriff, a detective asked the bank clerk, "Have you noticed anything specific about him?"

"Yes," the perturbed teller answered. "He seems to be better dressed every time he comes in."

● Did you hear the story about the prison bus that collided with the cement truck? The police are looking for hardened criminals.

● A group of counterfeiters got careless and printed up a batch of $15 bills. They tried to pass 'em off at Cousin Ethel's Feed-and-Seed and Country Store. They bought 30 cents of chewing tobacco. Cousin Ethel rang up the purchase and gave 'em back two $7 bills and two 35-cent pieces.

● A man went to apply for a job as deputy sheriff in the small town to which he had just moved. The sheriff said, "I'll have to ask you a few questions. Now, what days of the week start with 'T'?"

"Uh, today and tomorrow?" the newcomer responded.

PORK CHOP

"You'd better go home and study," the sheriff said, shaking his head.

The next day, the man came back. The sheriff asked this time, "Who killed Abe Lincoln?"

"I don't know," answered the man.

"Well, go home and find out," said the sheriff.

The man went home and walked in the door. "Did you get the job?" his wife asked anxiously.

"I think so!" he replied happily. "And they've already got me working on a murder case!"

- A California rancher was arrested for poaching, cooking and eating rare condor.

When the shocked game warden asked him what it tasted like, the rancher said, "Oh, it's a cross between a bald eagle and spotted owl."

- The farmer's wife asked her husband, "Why did you have the policeman put 80 mph on the ticket when he arrested you for driving 60 mph?"

Her husband replied, "I want to get a better selling price for the car when I sell it."

- A small town policeman directing traffic at a busy intersection tried to flag down a speeding station wagon as it approached him. The car kept on going. As it passed, the cop blew his whistle. The car sped on, so the officer jumped in his squad car and pursued the vehicle, his siren wailing and lights flashing.

After several blocks, the policeman got the car to pull over and found that the driver was a young woman accompanied by about a dozen children.

"Lady, don't you know when to stop?" the policeman demanded.

"Oh, my goodness, Officer," she replied sweetly. "These aren't all mine."

- "One thing I just don't understand," the judge said to the burglar standing before his bench. "Why did you break into the same store three nights in a row?"

"Your Honor, it's like this," the thief began. "I picked out a dress for my wife to wear to the Farm Progress Show and I had to change it twice."

- The town drunk was in court yet another time. The judge scolded, "I thought I told you I didn't want to see you in here ever again!"

The drunk slurred, "That's what I tried to tell the officer, but he wouldn't listen."

• "My good man," said the visiting chaplain to the prisoner in the penitentiary, "how did you happen to come to this sad place?"

"Well, sir," replied the convict, "you see in me the unhappy victim of the unlucky number 13."

"Really!" said the priest. "How was that?"

"12 jurors and one judge, sir," the prisoner whined.

• Two farmers were visiting over coffee at the town diner. One of them remarked, "Hey, did you hear about the local woman who was arrested the other day for sending out 1,500 anonymous valentines? Each beautifully crafted and steamily poetic valentine was perfumed and signed by hand with 'Guess who?'"

"So what?" asked his friend. "Why would anyone get arrested for sending anonymous valentines?"

"Well," replied the first farmer, "it was discovered she was a marketing-oriented divorce lawyer."

• When the police arrested a young medicine peddler for selling eternal youth pills, they discovered that he was one of those repeat offenders. Seems he had been arrested on the same charge in 1175, 1854 and 1914.

• Slippery Pete was on trial again for robbery and awaiting the jury's verdict. The lead juror announced the verdict, "Not guilty."

Rising to his feet, Pete announced in jubilation, "Great! Does that mean I get to keep the money?"

• A moonshiner was visited by a revenue agent.

"What's in the keg?" the suspicious agent asked.

"Nothing but water," the moonshiner said innocently.

The agent pried off the lid and took a sniff. "Why, that's moonshine liquor, old man!"

"What do you know?" the man gasped in feigned amazement. "The good Lord's done it again!"

• A shoplifter, caught stealing a necklace, pleaded with the jewelry store manager, "Please don't call the police. Can you find it in your heart to just let me pay for the necklace?"

The kind-hearted manager agreed to it and presented the bill to the thief. "Uh, that's a little more than I planned to spend," said the shoplifter. "Could you show me something a little less pricey?"

Chapter 35

Let Me Give You My Card

"I used to be a lawyer, but now I am a reformed character."

—Woodrow Wilson

● A defending attorney was cross-examining a coroner. The attorney asked, "Before you signed the death certificate, did you take the man's pulse?"

"No," the coroner replied.

Then the lawyer asked, "Did you listen for a heartbeat?"

"No," said the coroner.

"Did you check for breathing?"asked the lawyer.

"No," replied the coroner.

The attorney pushed on, "So when you signed the death certificate, you had not taken any steps to make sure the man was indeed dead, had you?"

The coroner, getting tired of the interrogation said: "Well, let me put it this way. The man's brain was sitting in a jar on my desk, but for all I know he could be out there practicing law somewhere."

● Two farmers were commiserating about today's rising lawyer fees. "I'm beginning to think that my lawyer is too interested in making money," grumbled Bruce.

"Why do you say that?" asked Burl.

"Well," Bruce explained, "just the other day I got a bill that stated: For waking up at night and thinking about your case: $150!"

● A new client had just come in to see a famous lawyer.

"Can you tell me how much you charge?" asked the client.

"Of course," the lawyer said. "I charge $200 to answer three questions."

"Well, that's a bit steep, isn't it?"

"Yes, I suppose," answered the lawyer. "And what is your third question?"

● Do you know what the most common name is that lawyers name their daughters? Sue.

And their sons? Bill.

● Why won't sharks attack lawyers? Professional courtesy.

● A farm hand wanted to go hunting but needed a dog, so he stopped at the local kennel. "All we have is ol' Judge," the breeder said. "And he's a prized purebred, so he'll cost $600."

"I don't have that kind of money," shrugged the hunter. "Will you rent him for the day?" The owner agreed.

A few hours later, the man returned blurting out praise for Judge. "He's the best dog in the whole world! Can't you come down on the price at all?"

But the owner wouldn't budge. So, the farm hand scrimped and saved until he gathered enough money. He returned to the kennel only to discover Judge's price reduced to $25. "What's going on?" the man complained. "Last time you wouldn't deal with me, and now he's almost free!"

"It's sad," the owner sighed. "Some other hunters took Judge out and began joking and calling her 'Lawyer.' Now all he does is sit around and whine."

● Two convicts in the rural county jail were passing time telling jokes. "Hey, Jake," Toby said, "what do you call a lawyer with an IQ under 50?"

Jake shrugged his shoulders and replied, "I dunno. What?"

"Your Honor!" Toby laughed out.

● A Mexican bandit was known for crossing the Rio Grande River from time to time and robbing Texas banks. Finally, a reward was offered for his capture, and an enterprising Texan decided to round him up.

After a lengthy search, the Texan traced the Mexican to his favorite watering hole, snuck up behind him, put his trusty six-shooter to the bandit's head and said, "You are under arrest. Tell me where you stashed the loot or I'll blow your head off!"

But the Mexican didn't speak English, and the Texan didn't speak Spanish. Luckily, a bilingual lawyer was in the saloon and acted as translator. The terrified bandit blurted out, in Spanish, that the loot was buried under the oak tree in back of the cantina.

"What did he say?" demanded the Texan.

The shifty lawyer answered, "He said, 'Forget you, jerk. You wouldn't dare shoot me.'"

● A lawyer opened the door of his BMW, when suddenly a car came along and hit the door, completely tearing it from the car. When the police arrived, the lawyer was complaining bitterly about the extensive damage to his precious BMW.

"Officer, look what they've done to my Beeeeemer!" he whined.

"You lawyers are so materialistic, you make me sick!" retorted the officer. "You're so worried about your stupid BMW, you didn't even notice that your left arm was ripped off."

"Oh, my God!" screamed the lawyer, looking down at his bloody left shoulder where his arm once was. "Where's my Rolex?!"

● What do you call an automobile accident between two lawyers? A Saab story.

● A dog ran into a country butcher shop and nabbed a roast off the counter. Fortunately for the butcher, he recognized the dog as his neighbor's. The neighbor happened to be a lawyer.

Incensed at the theft, the butcher called up the dog's owner and said, "Hey, if your dog stole a roast from my butcher shop, would you be liable for the cost of the meat?"

"Of course. How much was the roast?" answered the lawyer.

"$8.50."

A few days later, the butcher received a check in the mail for $8.50. Attached to it was an invoice that read: "Legal Consultation Service: $150."

● The lawyer's son wanted to follow in his father's foot steps, so he went to law school. He graduated with honors and returned home to join his father's firm.

At the end of his first day at work, the son burst into his father's office and exclaimed, "Father, in one day I broke the accident case that you've been working on for 10 years!"

"You idiot!" the father responded. "What do you think we've been living off all these years?"

● A 747 was having engine trouble, and the pilot instructed the cabin crew to tell the passengers to sit back down in their seats and prepare for a crash landing.

A few minutes later, the pilot asked his crew if everyone was buckled in and ready.

"All set back here, sir," came the reply, "except one lawyer who is still going around passing out his business cards."

● A junior partner in a Manhattan firm was sent to California to represent a long-term client accused of robbery. After days of trial, the case was won, and the client was acquitted and free to go. Excited about his success, the attorney typed a message into the firm's paging system: "Justice prevailed."

The senior partner replied in haste: "Appeal immediately."

● Taking his seat in his chambers, the judge faced the opposing lawyers. "So," he said, "I have been presented, by both of you, with a bribe." Both lawyers squirmed uncontrollably. "Mr. Franklin, you gave me $15,000. And you, Mr. Hansen, gave me $10,000."

The judge reached into his pocket and pulled out a check. He handed it to Mr. Franklin. "Now then, I'm returning $5,000, and we're going to decide this case solely on its merit."

● Lawyer: Someone who can write a 10,000-word document and call it a brief.

● Three law partners were cavorting in Miami Beach on vacation together.

All of a sudden, Norman shouted, "Oh no! I forgot to lock the safe in the office."

"So, what's the problem?" replied one of his partners. "We're all here, aren't we?"

● A farmer slapped down three $100 bills to pay the local attorney's $200 fee. Immediately, the attorney was faced with an ethics problem: Should he tell his partners about the extra $100?

● Contrary to Sunday school teachings, heaven and hell are actually located side by side. Not surprisingly, the devil (who was not a very good neighbor) often allowed the flames and heat to swell over to heaven's side. As a result, St. Peter approached the devil and politely asked him to build a wall along his boundary to control the heat and flames.

The next day, St. Peter was pleased to see that a 12-foot wall had been constructed—yet it was built far into heaven's property. St. Peter called the devil and demanded that the wall be moved back to the boundary line.

The devil defiantly shook his head and refused.

"Then you give me no choice but to sue you," St. Peter asserted.

A sinister smile appeared on the devil's face as he replied, "Is that so? And just where do you intend to find a lawyer?"

● A farmer and a lawyer were fishing in the Caribbean. "What brings you here?" the farmer asked his fellow fisherman.

"Everything I owned was destroyed by a fire. When the insurance company paid up, I decided to just keep the money. So, here I am," explained the lawyer.

"Wow, what an amazing coincidence," replied the farmer. "I'm here because my house and all by belongings were washed away by a flood. My insurance company paid for everything, too."

The lawyer looked confused. "Just between you and me," he said in a soft voice, "how did you start a flood?"

● A preacher and an attorney were talking one day about the mistakes they had made in their respective professions and how they dealt with them. The lawyer bragged that because he was a lawyer, if he ever made a really big mistake, he just shuffled a few papers around and pulled a few legal maneuvers and it was covered all up. If it was just a small mistake, he would just ignore it and go on with his life.

The attorney turned to the preacher and asked how he handled his big mistakes.

"Well," the preacher said, "if I make a really big mistake, I just ask the Lord for forgiveness."

The attorney replied, "But what about small mistakes? How do you handle them?"

The preacher answered, "Actually, just last Sunday, in my sermon I was quoting Jesus from the gospel of John, Chapter 8 where he said, 'Your father is the devil, and you do exactly what he wants. He has always been a murderer and a liar.' Instead, I said, 'He has always been a murderer and a lawyer.'"

Upon hearing this, the lawyer became indignant and retorted, "Well, how did you handle it?"

The preacher replied, "It was such a small mistake that I just ignored it and went on."

● The young, sharp attorney was cross-examining an elderly farmer who had witnessed an accident.

"You say you were about 35 feet from the scene of the accident? Sir, let me remind you that you are 86 years old. Just how far can you see clearly?"

"Well," the farmer coolly began, "When I wake up I see the sun, and they tell me that's about 93 million miles away."

● A very wealthy man, old and desperately ill, summoned to his bedside his three closest advisors: his doctor, his priest and his lawyer. "I don't know," the old man said. "They say you can't take it with you, but what if they are wrong? I'd like to have something with me, just in case. So, I am giving each of you an envelope containing $100,000, and I would be grateful if you would put the envelopes in my coffin at my funeral, so that if it turns out that it is useful, I'll have something to get by on."

The three men each agreed to carry out the dying man's wishes.

Sure enough, after just a few days, the old man passed away. At his funeral, each of the three advisors was seen slipping something into the coffin.

After the burial, as the three men walked away together, the doctor turned to the others and said, "Friends, I have a confession to make. As you know, there have been numerous cutbacks in funding for the hospital. Our CAT scan machine broke down and we haven't been able to afford a new one. So, I took $20,000 of our friend's money for a new machine and put the rest into the coffin as he asked."

At this, the priest admitted, "I, too, have a confession to make. As you are aware, our church is simply overwhelmed by the problem of the homeless. The needs keep increasing and we have nowhere to turn. I took $50,000 from the envelope for our homeless fund and put the rest in the coffin as our friend requested."

Fixing his gaze on the two men in front of him, the lawyer said, "I am astonished and deeply disappointed that you would treat so casually our friend's dying wish. I want you to know that I placed in his coffin my personal check for the full $100,000."

● The gray-haired attorney wrapped up his prosecution with a powerful, moving final statement. Attempting to deliver something as thought-provoking as his opponent's speech, the clever defense lawyer turned to the all-male jury and said, "Now, gentlemen, should this charming young lady be forced to spend her days in a lonely cell, or should she return to her beautiful little apartment at 226 Jackson Ave., telephone 555-7658?"

You Left It Where?

An old spinster, self-appointed supervisor of small town morals, accused a farmer of having snuck off in the middle of the afternoon to get drunk because, "with her own eyes," she had seen his wheelbarrow standing outside the local bar.

The accused farmer made no verbal defense, but the same evening he placed his wheelbarrow outside her door and left it there all night.

Chapter 36

If You're Ever Down in Texas...

"Ever since December 29, 1845, when Texas was kind enough to annex the USA, they have refused to put small in their dictionaries."

● "So, what did you get your wife for her birthday?" one Texan asked another.

"Oh, I got lazy this year and bought her something I wouldn't have to wrap," admitted the other. "I got her a yacht."

● Looking around a Texas ranch, the city man was startled to discover a large pond with several bell-ringing buoys floating on its wind swept surface.

He exclaimed to the rancher, "I thought buoys were a navigational aid!"

"They are," replied the Texan. "That's why we have 'em. This here is such a large ranch that the cows could never find the pond if they couldn't hear the ringing in them buoys."

"Oh, come on! You're pulling my leg, aren't you?" said the visitor.

"Pulling your leg?" the Texan said defensively. "Do you mean to tell me you never heard of cowbuoys?"

● A young lady from Kentucky was bragging about her state to a Texan. "In Kentucky," she said, "we have Fort Knox, where enough gold is stored to build a golden fence three-foot high completely around Texas."

"Go ahead and build it," the Texan drawled. "If I like it, I'll buy it."

● The bereaved widow went to a spiritualist to see if she could contact her dear departed husband.

The room was dark and the widow could hear what could be chimes as the spiritualist went into a trance. The chimes seemed to get louder, but they were off-key. Then she heard a voice.

"Mary," the voice said.

"Oh, Stanley, my dear," the widow said. "Is it really you?"

"It sure is," Stanley said.

"Are you all right, Stanley?"

"I'm fine, Mary, just fine."
"Is it nice there, dear?"
"Beautiful! Absolutely beautiful. Blue sky, pure air and green grass," Stanley replied.
"But what are those clanking noises?" asked Mary.
"Cow bells," Stanley said. "You've never seen such beautiful cows."
"Stanley, cows in heaven?"
"Who said anything about heaven?" Stanley said. "I'm a bull in Texas."

● One November afternoon, a Texan and his wife walked into a fine art gallery on Madison Avenue in New York. "How much is that picture in the window?" asked the woman.
"That, madam, is a Dufy—one of his early works—"
"Sure, sure," the Texan cut in. "How much ez it?" replied the store clerk.
"$22,000," replied the salesman.
"We'll take it," said the wife.
"Yes, sir! If you'll come this way," said the art dealer.
"Wait. How much ez this other paintin' on the wall?" continued the Texan.

EWE TURN

"That, sir, is a Vlaminick—," began the art dealer.

"Yeah, yeah. How much ez it?" interrupted the Texan.

"The Vlaminick is $43,000," said the dealer.

"Good," said the woman. "And that picture thair?"

So it went, until almost every painting on the walls had been bought by the debonair Texans. "Now, how many pictures does that make in all?" the Texan questioned.

"Let me see...15...20...25 works of art, madam," the wide-eyed salesman replied.

"That's fine, honey," said the Texan.

"Well, that takes care of them Christmas cards," the wife said. "Now we can start buyin' the presents."

● "Say, is that a new yacht?" Bob asked his Texan friend.

"Yeah, I had to buy myself a new one," replied the Texan.

"How come?" inquired Bob. "What happened to the old one?"

The Texan shook his head and said, "The darn thing got wet."

● Did you hear about the Texan who was traveling to Paris?

His wallet was classified as carry-on luggage.

● A Texan was traveling in Australia. "Don't you think that tower is beautiful?" asked his Australian guide.

"Well, now," drawled the Texan, "we've got towers as big as that or bigger back in Texas."

"What about this road?" asked the Aussie. "Have you ever seen any like it before?"

"Why sure, " said the visitor. "We've got lots of roads longer and wider than that."

They continued walking until they came to a field. Suddenly they saw a kangaroo hop by. "Well," said the Texan, "one thing I'll have to admit. Your grasshoppers are a bit larger than ours."

● Two Texans were driving their truck through the back roads of Texas when they came to an overpass with a sign which read, "Clearance: 11' 3"." The two got out and measured their rig, which was 12 feet and four inches tall.

"What do you think?" said one as they climbed back into the cab of the truck.

The driver looked to his left then to his right, checked his rear view mirrors, then shifted into first gear. "Not a cop in sight," he said. "Let's take a chance!" he said.

● Two Texans were swapping stories, trying to out do one another. "Did

I ever tell you about the time the local bank came to me to borrow money?" bragged the first.

"Oh, yeah?" drawled the second. "Well, did I ever tell you about the time I wrote a check and it bounced?"

"Now why would you admit that your check bounced?" the first Texan asked.

His friend replied, "Not the check, stupid! The bank!"

● Out in the far corners of Texas, the cowboys wear only one spur. They figure when one side of the horse starts running, the other side will, too!

● A wealthy Texan was helped out by his Congressman, who explained that his help was merely in the line of duty and that he would accept no remuneration.

The rancher persisted until the Congressman's secretary mentioned that golf was the lawmaker's favorite game, but that he needed new golf clubs.

Two weeks later, he received a message from the Texan: "Bought four clubs for you. I regret that only three of them have swimming pools."

● A Midwestern farmer was going to visit a ranch in Texas. In the airport, he got talking to a guy from Houston. "My wife made a millionaire out of me," the man remarked to the farmer.

Naturally, the farmer asked, "Well, what were you before?"

"A multimillionaire," groaned the Texan.

● A visiting Texan tipped the waiter in the country diner a $100 bill.

"I beg your pardon, sir," gasped the surprised waiter. "Do you realize how much you just gave me?"

"Yeah, well," began the Texan, "maybe next time you'll get my usual tip; that is if you bring me my food a little faster."

Nothing But Junk

A farmer doing business at the local bank was getting frustrated with his vain attempts at finding a working pen on the counter.

After a few minutes, the farmer threw up his hands and said loudly, "Where else but a bank would chain down a pen that doesn't write."

Keeping The Faith

Chapter 37

"Heaven goes by favor. If it went by merit, you would stay out and your dog would go in."

—Mark Twain

● "What do I have in my hand?" a preacher asked during a Sunday school lesson.

"A baseball!" a boy piped up.

"Good!" replied the preacher. "Now what does it make you think of?"

"God!" answered one little girl enthusiastically.

"Uh—why does a baseball remind you of the Lord?" asked the surprised clergyman.

"Well," explained the girl, "I know you didn't come over here to talk baseball."

● Little Timmy wanted a bike in the worst way. When he asked his dad for one, he replied, "Well, Timmy, if you're really good for a full two weeks, Jesus might just reward your good behavior with a new bike."

Taking his father's advice to heart, Timmy ran to his room to write a letter to Jesus. "Dear Jesus," he wrote, "if you'll give me a bike, I'll be good for two whole weeks."

After rereading his letter, Timmy rethought his terms and scratched out two weeks and wrote in one week.

This still didn't appeal to Timmy, and he again changed the amount of good behavior to two days. Still not happy with the deal, Timmy finally came up with the perfect solution.

He ran over to church, swiped the Mary statue, and ran home. This time he wrote a new letter: "Dear Jesus, if you ever want to see your mother again..."

● Wayne, a self-proclaimed atheist, died and was laid out at the local funeral home.

One of his cousins approached Wayne's casket, bowed her head and paid her respects. "Look at you, Wayne," the woman whispered. "All dressed up with no place to go."

● Little Nick was telling his grandmother what he had learned in Sunday school. "Adam was the first man," he said. "Methuselah was the oldest man, Job was the most patient man and Moses was the worst man."

"Nick, why would you say Moses was the worst man?" asked his grandmother.

Nick confidently replied, "Because he broke all Ten Commandments at once!"

● A small farm boy had behaved badly much of the day and had been in all kinds of mischief around the farm. When his exasperated father put him to bed that night and told him to say his prayers, the boy told him, "Please go away, Dad. I want to talk to God alone."

"What have you done that you don't want me to know about?" asked the father.

"If I tell you," the boy confessed, "you'll get angry, shout and probably punish me. But God will listen, forgive and forget about it!"

● In town to run an errand with her husband, the farm wife noticed a man loitering in front of the hardware store. Several men, she observed, stopped to talk with him and each gave him a little money.

He seemed so cheered by these encounters that she impulsively put $2 in an envelope, wrote "Godspeed" on it and took it over to him.

The next day, the woman made another trip to town to pick up materials for the farm. As she paid for her purchases, the woman was surprised to see the man rapping at the window to get her attention.

"Here's your $56, lady," he said cheerfully. "Godspeed won at 28 to one!"

● Bob worked constantly in his yard and had spectacular flowers and shrubs to show for his efforts.

A lady passing by one day said, "The Lord sure has given you a beautiful yard."

"Yes," said Bob, "but you should have seen what bad shape it was in when the Lord had it all to himself."

● A group of nuns was traveling through the countryside when their Buick ran out of gas. They spotted a filling station down the road, so they walked up to it, inquiring if they could borrow a gasoline can. Well, there wasn't any can, and the only container they could find was an old bedpan.

So, they filled it with gasoline, took it back to their car and were in the process of filling up their Buick when along comes old Sockless Sam in his 1975 Ford. He sees what's going on and comes to a dead stop. "What are you stoppin' for?" his cousin asks.

"I want to watch this," says Sam. "If that car cranks up and runs, I'm tradin' cars and switchin' churches!"

● Two little farm boys were debating about whether God actually exists.
"If God is for real how come we can never see him?" argued Norwin.
Dwight stared blankly into space, searching for the bit of wisdom that will finally end the silly argument. "All right, all right," Dwight finally said smugly. "If there is no God, Norwin, then who pops up the next Kleenex in the box?"

● What do you call a traveling nun? A roamin' Catholic.

● The Sunday school teacher of a little rural church was horrified when she saw the picture Frankie had drawn.
"Why, it looks like a cowboy walking into a saloon," she said.
"It is," answered Frankie proudly. "But don't worry. He's not going in there to drink any whiskey. He's just going in to shoot a man."

● Times have been hard for everyone. Just recently, a small Bible publishing company filed Chapter 11, Verses 1 through 14.

● A tattered $1 bill, in a bag on its way back to the U.S. Treasury, struck up a conversation with a worn out $20 bill. "I had a good life," the $20 bill said. "I traveled to nice stores, fancy restaurants and vacation hotspots. What about you?"
The $1 bill sighed, replying, "I went to church."

● Henry and Boyd, two old farmers, wondered if there was baseball in heaven. Sadly, Henry then suddenly died two days later. One night soon thereafter, his ghost came to visit Boyd.
"Boyd, old man, I have some good news and some bad news," said Henry. "The good news is, there's baseball in heaven. The bad news is, you're pitching tomorrow."

● Farmer Milo went to the doctor to see if there was anything that could be done about his snoring.
The doctor asked, "Does it bother your wife?"
Milo thought a moment and then said, "No, it just embarrasses her. It's the rest of the congregation it disturbs."

● "Please, God," the farmer prayed, "you know me. I am always praying to you, and yet nothing but bad crops, overdue bills and countless health problems have come my way. Look at the butcher next door. He's never prayed in his life, and he has nothing but prosperity, health and joy. How come a believer like me is always in trouble, and he is doing so well?"

A voice boomed from beyond, "Because the butcher doesn't bug me, that's why!"

● Jones cursed more than any other member of the congregation. The parson took him aside one Sunday and said, "Every time you swear, you must give $10 to the nearest stranger. Maybe that will cure you."

As Jones turned to walk away, he stubbed his toe and silently handed a $10 bill to a woman entering the building.

"OK," she whispered, "but can you wait till after the service?"

● A farm couple was leaving church after Sunday services. "Did you see the designer suit on the woman in front of us?" the wife asked. "And that hat on that woman across the aisle? And the frilly blue dress on the woman sitting to your left?"

"Well, no," the husband was quick to confess. "I'm afraid I dozed off."

She gave him a sharp look. "A lot of good church does you!"

● "Now why do you think Noah took two of every kind of animal into the Ark?" asked the teacher.

Johnny was ready with the answer: "Because he didn't believe that cockamamie story about the stork."

● As the Sunday school teacher was describing how Lot's wife looked back and turned into a pillar of salt, little Norman interrupted.

"My mother looked back once while she was driving," he announced. "And she turned into a telephone pole."

● One year a small farm community had a lengthy dry spell. The pastor scheduled a mass and asked all the people to come and pray for rain. The pastor knew of two elderly bachelor brothers who were not avid churchgoers and drove out personally to invite them to the gathering.

As it turned out the older brother did come to the service. After the service, the minister said to him, "It sure is good to see you here. But where is your brother?"

Caught off guard, the brother responded, "He stayed back to close the windows."

● The Sunday school teacher asked her young class how Noah spent his time on the Ark. "Do you suppose he did a lot of fishing?" she asked.

"I don't think so," replied a six-year-old, "he only had two worms."

● Unfortunately, children don't always get out of Sunday school what we would like them to. Here are some "facts" that a group of children learned in church school in England...

— Noah's wife was called Joan of Ark.
— The Fifth Commandment is Humor Thy Father and Thy Mother.
— Lot's wife was a pillar of salt by day, and a ball of fire at night.
— Salome was a woman who danced naked in front of Harrod's department store.
— Holy acrimony is another name for marriage.
— The pope lives in the vacuum.
— Paraffin is next in order after seraphim.
— The patron saint of travelers is St. Francis of the sea sick.
— Iran is the Bible of Moslems.
— A republican is a sinner mentioned in the Bible.
— The first commandment was when Eve told Adam to eat the apple.
— It is sometimes difficult to hear what is being said in church because the agnostics are so terrible.

● The Sunday school teacher asked, "How many of you want to go to heaven?"

Every hand went up but Mikey's. The teacher asked him, "Why, Mikey, don't you want to go to heaven?"

Mikey answered, "Not if this bunch is going!"

● A minister walked into the vestry and was shocked to discover his wife with both hands in the collection plate.

"Ethel!" he shouted, "what on earth are you doing?"

She said, "Looking for a button to sew on your coat."

● "Boy, sometimes it can feel like you've been in there forever!" Arthur complained as he was leaving the church.

"Yeah? And I think the whole congregation knows you feel that way!" the wife said, embarrassed by her husband's brashness.

"Why do you say that?" he asked.

She replied, "You seemed so bored in there, I think you fell asleep in the middle of your nap!"

● "Today's lesson, boys and girls," the teacher began, "teaches us about the influence of kings and queens. But there is a higher power. Who can tell me what it is?"

After a few moments, Sam called from the back: "Aces?"

● A young man had just entered his first country parish as a new priest and was given a task to perform: Rid the bell tower of pesky bats.

Father Adam enthusiastically went right to work and tried a variety of techniques to rid the bell tower of the bats. By the end of the week, he was very disgruntled as his efforts were in vain, and the bats had still not left.

Seeking advice, Father Adam went to the Monsignor. "What shall I do, Father George?" he said. "I'm afraid I have already failed at my first task at my new home."

The Monsignor, leaned over and replied in a soft voice, "Why don't we confirm them? Then they're bound to stay away!"

● Three Chinamen died and appeared before St. Peter at the Pearly Gates. "I'm sorry gentlemen," he said, "but only Christians are allowed into Heaven."

"Please," begged the Chinamen, "while we are not Christians, we still wish to enter Heaven."

St. Peter told them that if they each demonstrated knowledge of a Christian event, they would be allowed entry. The first was asked about Christmas. "Ah yes, Christmas is when kids dress up like monsters and eat turkey for hours." St. Peter shook his head.

The second was asked about Pentecost. "That when fat man in red suit light off firecrackers." Again, St. Peter shook his head.

The third was asked about Easter. "That when good man nailed to cross and killed," piped the third Chinaman. "Man buried in tomb. Three days later, rock moved away from tomb and man comes back to life."

St. Peter's eyes lit up and was about to allow entry into Heaven, when the Chinaman continued, "Then man sees shadow, goes back into tomb and winter last another six weeks."

● The minister was explaining the facts of life to his eight-year-old daughter. The young girl listened attentively as her father explained about the birds and the bees.

"Daddy," the little girl interrupted, "does God know about this?"

Poor Choice Of Words

A young priest, new with the prison system, was sent to console Mean Max, who was soon to see the electric chair.

As Max was being led to the chair, the flustered priest not wanting to say, "Goodbye" which sounded terribly final, or "See you later," which sounded too casual, finally said the condemned man, "More power to you!"

Chapter 38

God's Spokesmen

"Actually that's our jobs as ministers. When we're not comforting the afflicted, we should be afflicting the comfortable."

● Farmer Ollie was near death, and his family summoned a priest to perform last rites. As his condition quickly worsened, he motioned frantically for something to write on, so the priest handed him his pen. With his last bit of energy, Ollie scribbled a note, then died. In the commotion, the priest put the unread note in his pocket.

As the priest was giving the eulogy at Ollie's funeral, he realized he was wearing the same jacket as when Ollie died. He told the congregation, "You know, ol' Ollie handed me a note just before he passed on. I haven't looked at it yet, but knowing Ollie, I'm sure there's a word of inspiration for us all."

The priest opened the note and read aloud, "You're standing on my oxygen tube!"

● Just before the minister was to deliver his evening sermon, an usher handed him a note. The preacher announced someone had left a car locked with the lights on in the parking lot.

"The implication seems to be," he added wryly, "that the battery may run down before I do."

● The rural minister parked his car in a no-parking zone during a trip to the big city and attached the following message to his window shield: "I have circled this block 10 times. I have an appointment to keep. Forgive us our trespasses."

When he returned, to his car he found this reply attached to his own note along with a ticket: "I've circled this block for 10 years. If I don't give you a ticket, I lose my job. Lead us not into temptation."

● "There are 98 bars in this town and I've never been in one of them!" the country preacher cried.

From the back row, a little voice piped up, "Which one is that?"

● One day while driving his pickup back from town through a thunderstorm, a farmer got a flat tire just outside of a monastery. A monk came out and invited the farmer inside to have dinner and spend the night. The farmer graciously agreed.

That night, he had a wonderful dinner of fish and chips. He decided to compliment the chef. Entering the kitchen, the farmer asked the cook, "Are you the fish fryer?"

"No," the man replied with a smile, "I'm the chip monk."

- The new minister knocked on the door of a farmhouse.

"Is that you, angel?" called a woman's voice.

"No," came the answer, "but I'm from the same department."

- After watching a minister match coins with a member to see who would pay for coffee, a bystander asked, "Preacher, doesn't that constitute gambling?"

The preacher replied, "Not at all, son. It's simply a scientific method of determining just who is going to commit an act of charity."

- One Sunday, the minister of a small country church told the entire congregation to stand during the offertory. He instructed everyone to reach forward to the person standing in front of them and take their purse or wallet.

Then he added, "Now open their pocketbook and give as you always wanted to, but felt you couldn't afford!"

- Our country minister says he does not mind a man looking at his watch during one of his sermons, but he does resent his shaking it to find out if it is still running.

- Preacher Jones was fond of peach brandy. One of his vestrymen said to him, "I'll be happy to give you a bottle of peach brandy if you'll promise to acknowledge it in the church bulletin. The preacher thought about it for a few moments and agreed.

In due time, a note in small print appeared in the bulletin: "The rector wishes to thank the vestryman for his recent gift of fruit...and the spirit in which it was given!"

- A cannibal chief had trouble with a choice meal cooking on his newfangled broiler and asked a witch doctor for advice. "Hmm, does he have a round white collar and a black robe?" asked the witch doctor.

"No," answered the chief, "he has a shaved head with a circle of hair, a long brown cassock and sandals."

"Oh, no wonder," laughed the witch doctor. "He's not a broiler, he's a friar."

● The newly hired pastor was making a lot of changes, and the budget began to swell. The finance chairman decided it was time for a talk. "I see you've hired a man to care for the church grounds," the chairman began. "Did you know the previous pastor used to do that job?"

"Oh, I'm aware of that," the new pastor answered, "But he said he doesn't want to do it anymore."

● Two pastors were talking as they strolled around the park. "Hey, Joseph," said one, "what happens when you don't pay your exorcist's bill?"

"I don't know, Daniel," the other replied. "The devil makes you do it?"

Daniel replied, "No, you get repossessed!"

● A young priest was flying home to visit his parents, when the 747 experienced engine failure. The pilot, struggling with the controls, managed to straighten out the plane and then asked the flight attendants if the passengers were overly nervous.

"They're near the point of panic," she reported.

The pilot turned the controls over to his co-pilot and walked back to where the priest was sitting.

"The rest of the passengers are alarmed," he said in an urgent tone. "Please, do something religious."

So the priest took up a collection.

● A preacher was several minutes into his sermon when a farm hand in the back cried out, "I can't hear you!"

"Really?" answered an irritated man sitting in front, "I'll trade places with you!"

● The visiting Reverend, preaching in the small town congregation of a friend, stopped in the middle of his long, elaborate sermon and signaled to the sexton. "In the second row," he whispered, "there is a man who is absolutely asleep...snoring! Go wake him up."

"Me?" frowned the sexton. "Is that fair?"

"What do you mean, 'is that fair?'" asked the perturbed Reverend.

"Well, you put him to sleep," responded the sexton sheepishly. "So, I think you should be the one to wake him up."

● A visiting country clergyman delivered an entertaining talk at a small town banquet. The following day, he was to address the Women's Club in a neighboring village. Wishing to repeat some of his stories at the latter affair, he asked the reporter at the banquet not to print any of his anecdotes in the paper.

The next day, the account of the clergyman's speech concluded with: "Rev. Smith also told a number of stories that cannot be published."

● "I must warn you," the merry new preacher told his congregation, "when all is said and done, I have a tendency to keep right on talking."

● "Actually, there's a lot to be said for sin," one clergyman said to another. "After all, without it, we'd be out of a job."

● During his last visit to the U.S., the pope was so impressed with the limousine provided for him that he asked if he could drive it. The startled chauffeur balked at the idea, but the pope was so insistent, they finally pulled over and switched seats. The pope immediately stomped on the gas pedal and began weaving boldly through the rush-hour traffic.

After only a few minutes, they were spotted by a patrolman and pulled over. After a brief warning, the officer meekly returned to his partner. "Forget about this one," he mumbled, shaking his head. "I couldn't give him a ticket."

"Why? Is it the mayor?" asked the partner.

"Bigger," said the wide-eyed cop ominously.

"Not the president, surely?" asked his partner.

"Look," said the first officer, "I don't really know who he was, but his chauffeur is the pope."

● The rural farm wife said to her young son, "I'm going out to collect eggs. If the butcher comes, let me know. I need to talk to him."

A few minutes later, the minister stopped by. The boy, forgetting whom his mother wanted to talk to, called out back, "Ma! That man's here now."

His mother called back, "I can't come in quite yet. Give him a dollar out of my purse and tell him we didn't like his tongue last week. And if it's no better this week, we're gonna change."

● A stranger entered the church in the middle of the sermon and seated himself in the back pew. After a while, he began to fidget. Leaning over to the white-haired man at his side, evidently an old member of the congregation, he whispered, "How long has he been preaching?"

"Thirty or forty years, I think," the old man answered.

"I'll stay then," decided the newcomer. "He must be nearly through."

Buy Both Books...Save $8!

Regardless of the season, ***Country Chuckles, Cracks & Knee-Slappers*** and ***More Country Chuckles, Cracks & Knee-Slappers*** offer the very best in gifts for family members, good friends or even that special person who moved to the city but still has his or her heart or roots out there in rural America.

Whether you give these folks both books or split your order into two special gifts, you'll **save $8** by ordering both great country humor books at one time. *Country Chuckles, Cracks and Knee Slappers* sells for $11.95 plus $4 shipping and handling. The new joke book, *More Country Chuckles, Cracks and Knee Slappers* sells for $12.95 plus $4 shipping and handling.

If you buy both books (a total value of $32.90 for 464 pages of jokes, which works out to just a little more than 1 cent per joke), we'll waive all of the shipping and handling charges...**saving you $8** on your order. Just mention Priority Code LP97 to get these savings! Or you can also order the original book at $15.95 each (including shipping and handling) or the new book at $16.95 each (including shipping and handling).

So write or call today while you're still thinking about it...there's no easier way to get those troublesome and time-consuming gift buying chores out of the way right now than by ordering one or both of these books today.

"We ordered *Country Chuckles, Cracks And Knee-Slappers* and liked it so well that we have ordered more copies to give as Christmas presents."

—Nancy Pullman, Lifton, Ind.

Send your check (U.S. funds only) or credit card authorization (MasterCard, Visa or American Express) for these books to:

Lessiter Publications, P.O. Box 624, Brookfield, WI 53008-0624.
Telephone: (800) 645-8455 in the U.S. or (414) 782-4480.
Fax: (414) 782-1252 E-mail: info@lesspub.com

Priority Code: LP97

You Need The Original, Too...

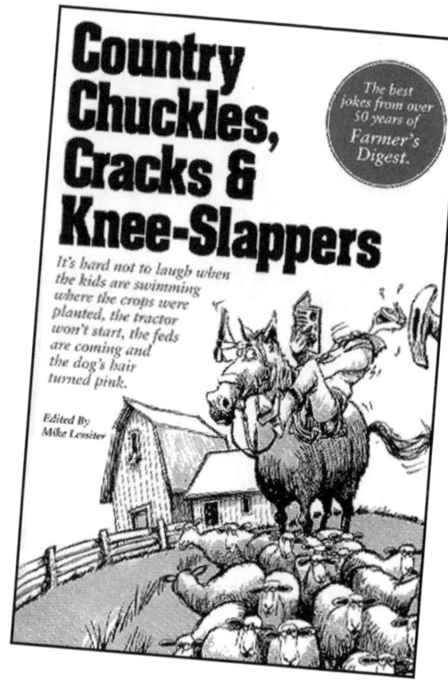

Get ready for a wild and crazy laugh-a-minute ride down the back roads of America through the pages of this 256-page book—the original country joke book produced several years ago. ***Country Chuckles, Cracks & Knee-Slappers*** includes 46 giggle-inducing chapters that will knock you off your rocker with 1,241 of the funniest jokes you'll find anywhere. It's the perfect companion to the book you're holding in your hands and available for **only $11.95** per copy (plus $4.00 for shipping and handling). By buying the original country humor book and enjoying it along with this one, you'll have 2,328 of the very best jokes to ever come out of rural America!

"Inside this laugh-packed book are 1,241 of the very best gut-busting jokes selected from hundreds of back issues of *Farmer's Digest* magazine going back over 50 years. This book treats readers to the greatest humor ever found as print on paper..."

—*The Midwest Book Review*

This book sells for $11.95 plus $4.00 for shipping and handling. Send your check (U.S. funds only) or credit card authorization (MasterCard, Visa or American Express) for these books to:

Lessiter Publications, P.O. Box 624, Brookfield, WI 53008-0624.
Telephone: (800) 645-8455 in the U.S. or (414) 782-4480.
Fax: (414) 782-1252 E-mail: info@lesspub.com

Ask us for our newest book catalog to check out the more than 85 books, Special Management Reports and publications that we offer to our readers at reasonable prices.